PEOPLE,
THEIR NEEDS, ENVIRONMENT,
ECOLOGY

ESSAYS IN
SOCIAL BIOLOGY

Volume I

PRENTICE-HALL INTERNATIONAL, INC., *London*
PRENTICE-HALL OF AUSTRALIA, PTY. LTD., *Sydney*
PRENTICE-HALL OF CANADA, LTD., *Toronto*
PRENTICE-HALL OF INDIA PRIVATE LIMITED, *New Delhi*
PRENTICE-HALL OF JAPAN, INC., *Tokyo*

PEOPLE,
THEIR NEEDS, ENVIRONMENT,
ECOLOGY

ESSAYS IN
SOCIAL BIOLOGY
Volume I

Bruce Wallace

*Division of Biological Sciences
Cornell University*

PRENTICE-HALL, INC.
Englewood Cliffs, New Jersey

ISBN   0-13-656827-0   0-13-656835-1
Library of Congress Catalogue Card Number 79-167789
Printed in the United States of America

71   18993

To   MCW
     DBW
     RSW

Undoubtedly the most momentous piece of progress of 1969 occurred in Dijon, France, on August 14 when Neil Rappaport of Festoona, N.J., lost forever the blueprints for a machine he had just invented which would have made it possible to create a louder noise than had ever been heard before.

Russell Baker, The New York Times, February 24, 1970. ©1970 by The New York Times Company. Reprinted by permission.

Acknowledgments

The author wishes to acknowledge that these essays owe their existence to a criticism made by Professor Jacques Barzun of college-level "science survey" courses. Whether I have succeeded in meeting this criticism is, of course, another matter.

A large number of colleagues at Cornell University have helped me in one way or another during the preparation of this book. Dr. Richard B. Root outlined the fascinating aspects of ecology which relate to the selection from *Iberia*. Similarly, Dr. Thomas Eisner helped me to identify many points of interest concerning communication. Dr. Richard D. O'Brien aided me in understanding the details of the transmission of nerve impulses and the modification of this transmission by chemical means. In many conversations, Dr. Gerald R. Fink explained matters related to diseases, their transmission, and the action of antibiotics. Indeed, virtually all persons who have taken part in the *Biology and Society* lectures at Cornell University have aided me immensely.

No less than my colleagues on the faculty, I must also thank the students of *Biology and Society* who undertook to read and criticize the essays during their preparation. It is a pleasure to acknowledge with special thanks the efforts of Cynthia Ravitski, Terri Schwartz, and Mark Zabek. Penny Farrow and Florence Robinson also had numerous encouraging comments and useful criticisms during this same period.

Within my own family, each member read (and seemingly enjoyed) one or more essays. Considered with my family are a host of college students who were unexpectedly pressed into reading these essays while visiting our home – both in Ithaca and abroad during the summer of 1970. Every reaction whether heeded or not has helped in the preparation of this book. And each and every one has been appreciated.

Finally, I must thank Drs. Franklin A. Long and Raymond Bowers of the Program for Science, Technology, and Society at Cornell University who arranged for a release of two months from my regular duties so that I could complete these essays. I wish to thank Dr. Harry T. Stinson for his efforts in consummating these arrangements. Financial support during these months was provided through a grant from the Alfred P. Sloan Foundation. The Program for Science, Technology, and Society has been instrumental, as well, in supporting the *Biology and Society* lectures at Cornell, a series of talks on topics similar to those discussed in my own essays.

Contents

Contents

Preface

This book is like no other biology text I know. I say this immediately to warn the reader, both my professional colleagues and the student. These essays have been written to be enjoyed. I assume that the reader has had a modern high school biology course, that he enjoys reading, and that his instructor will fill in or enlarge upon background factual material wherever necessary. Proceeding from these assumptions, I have attempted to explore contemporary biology by means of single-topic essays – touching upon a variety of questions in as informal and nontechnical a manner as possible.

From classroom experience I know that the student's initial reaction to this text is one of discomfort. Where, he asks, is the glossary of terms from which routine quizzes are prepared? What can I memorize? What am I expected to learn? What will the professor ask on his examinations? These are the conditioned reflexes of today's student – reflexes honed to a fine edge during many years of secondary and higher education. Because of the exaggerated emphasis that is placed by schools on measurable performance, much of the student's commitment to any knowledge that he might have gained in class is shed as he hands in his final exam; at that moment he has passed or failed in his effort to please his professor. Next comes the pleasure of forgetting most of the unpleasant ordeal.

Today's serious problems are not in the classroom; they are outside. These problems demand an aroused public equipped with all the information and intellectual skill that individuals can muster. We can no longer afford to shuck our knowledge at the classroom door as we depart. My plea to both the student and the instructor, therefore, is to abandon the weekly vocabulary drill and the multiple-choice examination; instead, read the essays and their accompanying selections, enjoy them, react to them, and through prolonged discussion criticize and rethink them. If enjoyment and critical discussion cause a student to retain some of what he has learned, fine. We must by all means encourage him to enjoy biology, for it is only because he may later make use of what he has retained that we teach biology to the nonbiologist.

Before World War II, bird watching and nature study made an adequate biological background for someone whose daily efforts were spent in court, in a business office, in a factory, or teaching one of the humanities in high school or college. Biology of this sort offered a pleasant diversion to the business of earning a living. Simple biology, however, is no longer sufficient. Today, many of man's problems have a biological basis – numbers of persons, the production of food to feed the hungry, man's extermination of other living things, and the wholesale destruction of the environment by man's industry. None of us can afford any longer to be a mere spectator; where a few fight for immediate profit, the many must fight for a lasting place to live. Many problems — racial, genetic, eugenic — lurking on the horizon are also biological ones. It is ridiculous to expect intelligent action to come from biologically naive persons who, with fatal optimism, seek at most ad hoc solutions to grave problems. The decisions affecting irreversible governmental and industrial actions are largely scientific in nature; we can no longer be satisfied with science survey courses that merely train dillettante bird watchers and butterfly collectors. These courses must teach students to grapple with the serious problems of today — and tomorrow.

What precisely is the aim of a required course in elementary biology at the college level? What is it that the student should learn? Opinions differ. I believe that today's college student most likely has had a modern course in high school biology comparable to those prepared by the Biological Sciences Curriculum Study. I feel that he should understand the workings of his body and mind ("know thyself" is still an apt phrase) and how both continue to work under a barrage of external and internal challenges. He should see how mankind fits into the skein of life on earth and appreciate the constraints placed upon man if both he and the skein are to continue their existence. We are, as the astronauts said, confined to a raft that is a small blue gem floating in space; our problem is to keep the raft gemlike and blue. Finally, the student should develop a sense of understanding and a method of intellectual attack appropriate for the problems that confront (haunt?) us in our daily relationship with government, business, community, and neighborhood affairs. The continual reliance upon expedient solutions to grave problems resembles the instant-by-instant reflexive responses of a frightened hare attempting to elude a dog; it does not become the intellectual powers of man. A college-level biology course for nonmajors should help the individual visualize and evaluate probable outcomes of certain courses of action. It should teach him to cope with and refute the oldest and most seductive of all ad hoc arguments: "It smells like money."

How does one obtain these goals? In truth, no one knows. To ask each student to make personal observations on all matters of import is patently foolish; in our complex society such observations must be made vicariously. Consequently, in the sections that follow I have drawn upon the observations and experiences made and described not by scientists but by essayists, novelists, and historians. I have turned to these persons because they are not only professional observers but also are competent writers. In doing this I may have risked scientific precision but I have gained tremendously in other respects. A clinical account, no matter how accurate, cannot match Voltaire when he writes "a beggar all covered with sores, dead-eyed, the end of his nose eaten away, his mouth deformed, his teeth black, tormented with a violent cough . . . ."

Because students are initially discomforted by my approach, I want to explain further my use of literary selections in a science text. In my view they substitute for personal observations but not for the precise and structured observations of a field trip. They more nearly resemble series of strolls, some brief and others perhaps overly long. Each establishes a mood and offers an excuse to embark upon a series of discussions; the essays serve this latter purpose. No attempt has been made to write an essay about every possible topic posed by its accompanying selection nor have the essays been restricted to topics included within the selection. If the selections can be compared with strolls, the essays can be compared with the conversational ramblings of a talkative companion – one inspired by events he encounters but whose conversation is neither restricted to nor necessarily evoked by the passing scene.

The reader will find in these three volumes selections from a number of sources as well as a variety of single-topic essays. Their arrangement, of necessity, is not rigorously sequential. Sometimes in earlier essays I have omitted details needed to understand later ones. Sometimes I have repeated information. In any case, the effect is to make the essays just the more chatty and repetitious. When the reader has finished the last essay, he will be totally uninformed of such things as the difference between *polyploidy, polyteny,* and *polysomy,* but why shouldn't he be? There is a remote chance that he will have enjoyed himself at some point. If his enjoyment causes him to retain a knack for wrestling with today's problems, and if he uses this knack by speaking up publicly on controversial environmental and population issues, his enjoyment could prove to be the lasting contribution of the book.

Bruce Wallace

A Word to the Teacher

The use of this text will call for techniques and attitudes unlike those found in most biology classes. Student interest in the types of topics discussed in these essays is nearly overwhelming. Lecturers in the initial *Biology and Society* series at Cornell University (Biology 201-202, 2 points per semester) found themselves addressing sustained weekly audiences of 1,000 (students and townspeople) rather than the 25 or 30 that most had expected. The problems that arise from an enthusiastic response of this kind are of three sorts: (1) the handling of large numbers of students in an intimate way; (2) the choice of factual material to be stressed in the course; (3) the basis of grading students who enroll for credit.

How can large numbers of students be handled where the burden of a social biology course falls on one staff member? I suggest that this man (presumably a biologist) be conversant with the material covered by any one of a number of excellent college-level biology textbooks and, further, with the additional material to be found in some 130 *Scientific American* offprints. Each lecture should be restricted to no more than 20 minutes; it should present only the facts that are essential for understanding the assigned essay(s). The remaining 30 minutes should be reserved for student discussion and questions from the floor. The role of the lecturer during this period is to encourage student participation and to supply factual information when necessary in order to keep the discussion accurate. Where information is lacking or when someone questions the accuracy of an essay, a team of two or three students should be told to recheck the facts and return in a week's time with a five- or ten-minute report for the class.

Facts and the collection of factual material should be recurrent themes throughout all discussions. The types of facts, however, that are important in a type of course like *Biology and Society* need not be those that are considered important in most beginning biology courses. Vocabulary is trivial, except to the biology major whose livelihood will depend upon knowing it. Many of the details of ordinary biology courses are also trivial — especially those details that are rapidly forgotten by most students and, consequently, are so useful as test items. The facts that I feel are important are the obvious ones that tend to be neglected. These are the facts that at times are embarrassing to teach because the students react by saying, "Of course." They must be repeated again and again nevertheless. If simple facts were understood by everyone, permanent equipment would not be designed to wear out nor would disposable articles be made so often of indestructible materials.

My personal feelings on the matter of grades and grading are rather strong ones. The student who attends a social biology course and later, because of it, takes an active and intelligent part in community affairs attests to the success of the course and its teacher; the student who memorizes facts, talks in class, but refuses to become involved in society's affairs represents a failure. I have argued that grades for Cornell's *Biology and Society* should be limited to S/U (satisfactory-unsatisfactory) *only* and that those who take the course should receive credit only for use in accumulating general credits. This argument is based in part on my feeling that the choice of lecture topics and the overall format of the course can most likely remain flexible and stimulating only if traditional administrative groups and committees are kept uninvolved; once a standing committee feels that the course content falls within its domain, freedom to arrange novel discussions or sequences of off-beat lecture topics may disappear. In addition, however, I firmly believe that no appropriate grade in social biology can be given until at least two decades after the student's graduation.

Prologue

Dear Bruce and Roberta,

This has been a terrifying cruise. I will be thankful when we finally reach port. Wherever the ship first puts ashore, I will get off and fly home.

You would not believe how things have gone or what has happened. Remember the story about the woman who reported the lost elephant to the police sergeant? "This animal is pulling up my cabbages with its tail and you wouldn't believe what he does with them!" That is the way this trip has been — unbelievably absurd!

So few passengers had signed up by the time we sailed that we thought the cruise might be cancelled. All that is long past, but let me start at the beginning. Let me start with the Game! Immediately after we left the pier someone started this game of Monopoly. It turned out to be Monopoly *for keeps*. For keeps, believe me! When the bank first ran out of play money, someone suggested that the game could continue if we took parts of the ship — glasses, door knobs, faucet handles, and such things. Values were set by supply and demand; a perspiring man, you know, will pay a good deal to recover a missing shower head. The crew was tolerant; the purser, in fact, would reassign a passenger to a fresh room if the other players had taken either the lock or the ventilator from his cabin door. He would also reassign the passenger — for a price, that is — if the passenger himself had gambled and lost too many knickknacks from within his room.

Although we have never touched land since leaving the pier, the ship has rendezvoused on occasion with smaller boats and these have transferred passengers on board. At first this happened only infrequently, but the rate at which new passengers are put aboard has gone up tremendously. I am convinced that the rate at which they come aboard increases hand-in-hand with the number already aboard. They now greatly outnumber the original passengers, and some rather recent arrivals complain about the large number of even more recent newcomers.

We no longer have empty staterooms aboard. For the most part we must share a room with several others. The rooms are in pretty bad shape, of course, because of the Game. The latest craze has been to take the glass discs from the portholes. Most of these have been stolen for "Easter Island" cash and have been replaced by wood. It is impossible to look out but, as the newcomers say, "There's only water out there anyhow."

The ship is becoming ill-kempt; the filth is very bad. There are far too

1

many persons aboard. Furthermore, the facilities have been impaired by constant thievery. The rest rooms are no longer adequate. Impatient persons long ago began relieving themselves in dark passageways and obscure corners. Now, of course, it is necessary to do this regularly because of the number of persons aboard, and so these passageways and corners are officially designated latrine areas. I have recently noticed, however, that the more impatient passengers, because of the increased congestion, are relieving themselves quite publicly. Even well-lighted and well-traveled areas have become nasty underfoot.

We had quite a scare the other day — a terrible experience. One of our more enthusiastic and successful players found a metal plug protruding from the wall and, after a great deal of effort, pulled it out for use in the Game. To everyone's horror, water poured in. It seems as if we are now down to the hull itself. Many passengers are irate and have demanded that the hull be declared "off limits." The man responsible, however, bought a full-page ad in the ship's paper to explain glibly that the hull has now been reinforced with patching plates and, as a result, is actually stronger than it was when we sailed. This statement has reassured many. We all gratefully contributed to a fund for the sailors who risked their lives getting the patches in place. Other ardent players have pointed out, in the meantime, that these plugs are now essential to the Game, and if it is to continue, they must be removed whenever and wherever they are found. These players admit that there is a slight risk that the ship might sink, but they insist that the make-shift patches, though unsightly, are not dangerous. They seem quite confident that nearly any hole in the hull can be successfully repaired. In reference to the danger of the ship's going down, they say, "Not one chance in a billion."

You can see how things have gone and why I say, "What a cruise!" The fun is long gone. The Game rolls on at an ever-increasing pace. The small boats swarm about us now in great numbers with more and more passengers to be put aboard. The filth piles up. There is a chill that comes from thinking how things are going aboard this ship and what the outcome may be. Some passengers have demanded that the captain stop the ship; they want to get off. Those caught up in the frenzy of the Game, sensing the animosity of many that is aimed at them, have resorted to garish signs reading, "Love It or Leave It."

At any rate, I shall write again when I get a chance. In the meantime, if I thought anyone would or could believe it, I would write a book....

Love to all,

Daddy

SECTION ONE

People
and Their Needs

Introduction

Man's first concern has been and should continue to be the improvement of man's lot. When there were few of us, we could at times improve our lot by false means — by stealing from a world and an environment that seemed inexhaustible. Now there are many more of us. To make large numbers of us happy or, in times of war, to make large numbers of us unhappy calls for a tremendous destruction of the natural resources of the world. The production and distribution of food on the scale needed to feed mankind requires space, water, and energy; when the number of persons on earth is reckoned by the billions, these requirements become colossal and approach the limits of earth's ability to provide. The time for husbanding essential resources has arrived.

The hands of the clock cannot be turned back. How often have we heard this "justification" for progress, for forging ahead, and how often have we forged deeper into a morass? How often have we struggled with false solutions for problems that, unless they are confronted directly and honestly, have no solutions? Highways cannot relieve traffic jams if cars are put together faster than roads are laid. Larger and larger airports located ever farther from the cities they are intended to serve will not solve air-traffic problems if the number of airplanes increases faster than runways and if airport-to-city surface traffic is at a standstill.

What sorts of nontechnical readings bear appropriately on the numbers and needs of people? I am not sure that I have come up with adequate choices. A number of thoughts have occurred to me but one by one I have rejected them. Herodotus, for example, described the mighty force brought from Asia by Xerxes to destroy Athens:*

> As far as this point then, and on land, as far as Thermopylae, the armament of Xerxes had been free from mischance; and the numbers were still, according to my reckoning, of the following amount. First there was the ancient complement of the twelve hundred and seven vessels which came with the king from Asia — the contingents of the nations severally — amounting, if we allow to each ship a crew of two hundred men, to 241,400. Each of these vessels had on board, besides native soldiers, thirty fighting men, who were either Persians, Medes, or Sacans; which gives an addition of 36,210. To these two numbers I shall further add the crews of the pentecoters; which may be reckoned, one with another, as fourscore men each. Of such vessels there were, as I said before, three thousand; and the men on board them

accordingly would be 240,000. This was the sea force brought by the king from Asia; and it amounted in all to 517,610 men. The number of the foot soldiers was 1,700,000; that of the horsemen 80,000; to which must be added the Arabs who rode on camels, and the Libyans who fought in chariots, whom I reckon at 20,000. The whole number, therefore, of the land and sea forces added together amounts to 2,317,610 men. Such was the force brought from Asia, without including the camp followers, or taking any account of the provision-ships and the men whom they had on board.

To the amount thus reached we have still to add the forces gathered in Europe, concerning which I can only speak from conjecture. The Greeks dwelling in Thrace, and in the islands off the coast of Thrace, furnished to the fleet one hundred and twenty ships; the crews of which would amount to 24,000 men. Besides these, footmen were furnished by the Thracians, the Paeonians, the Aeordians, the Bottiaeans, by the Chalcidian tribes, by the Brygians, the Pierians, the Macedonians, the Perrhoebians, the Aenianians, the Dolopians, the Magnesians, the Achaeans, and by all the dwellers upon the Thracian seaboard; and the forces of these nations amounted, I believe, to three hundred thousand men. These numbers, added to those of the force which came out of Asia, make the sum of fighting men 2,641,610.

Such then being the number of fighting men, it is my belief that the attendants who followed the camp, together with the crews of the corn ships, and of the other craft accompanying the army, made up an amount rather above than below that of the fighting men. However I will not reckon them as either fewer or more, but take them at an equal number. We have therefore to add to the sum already reached an exactly equal number. This will give 5,283,220 as the whole number of men brought by Xerxes, the son of Darius, as far as Sepias and Thermopylae.

Such then was the amount of the entire host of Xerxes. As for the number of women who ground the corn, of the concubines, and the eunuchs, no one can give any sure account of it; nor can the baggage horses and other beasts of burden, nor the Indian hounds which followed the army, be calculated, by reason of their multitude. Hence I am not at all surprised that the water of the rivers was found too scant for the army in some instances; rather it is a marvel to me how the provisions did not fail, when the numbers were so great. For I find on calculation that if each man consumed no more than a choenix of corn a day, there must have been used daily by the army about 170,000 bushels, and this without counting what was eaten by the women, the eunuchs, the beasts of burden, and the hounds. Among all this multitude of men there was not one who, for beauty and stature, deserved more than Xerxes himself to wield so vast a power....

The Persian army was a tremendous one indeed. That rivers should be emptied in providing for the thirst of these men and their beasts is impressive. But only 5 million men were involved, a number surpassed now by at least a dozen of the world's largest cities. Today, any medium-sized city has water requirements – for drinking, sanitation, and industrial uses – that are met only by torrential rivers. Persons throughout the world must now be reckoned in billions, not millions as by Herodotus. The amount of water required per person today can be staggering.

The three selections included here were chosen for a variety of reasons. At the root of all other problems facing mankind is one that *must* be solved before rational solutions can be proposed for any other: *numbers of persons.* Man is a large mammal (not a colossus, to be sure, but a much larger-than-average mammal) and so his demands for water and for air are likewise great. Despite these substantial individual needs, man has multiplied until he has attained population numbers generally reached only by animals that are much smaller in size. Perhaps you remember, especially if you are middle-aged or from a rural area, the endless times you have walked about the house swatting flies. Now, imagine exterminating house flies with a fly swatter! The mainland Chinese have reportedly done just that; how better can one illustrate the incongruous numbers human beings have now attained? Pitted evenly against flies, and man wins!

Man is a large mammal whose numbers challenge those of much smaller ones. The maintenance of these enormous numbers over extended periods of time, the demands made on the environment merely to sustain these numbers, and the further absurd demands through the irrational use of resources for trivial ends (a government official has seriously discussed the future establishment of a three-hour Ceylon-United States postal service by rocket as if it were both necessary and desirable) make the encyclical of Pope Paul VI an important document of our time. By opposing artificial birth control the encyclical condones, even though it may not overtly encourage, population growth at its present rate. Consequently, one must interpret this document as lacking concern not for individual human beings, because the Church is compassionate, but for total *numbers* of individuals and their combined needs. Apparently, human beings in 9- and 10-digit quantities do not unduly alarm Roman Catholic authorities.

In one of the public discussions by the Church on artificial birth control, the tender phrase "invite the unborn to the banquet of life" (or a phrase very much like it) was used. The second selection, in bitterly satiric style, describes a "banquet of life" that might have stemmed the population explosion and poverty of 18th-century Ireland. The specifics of Swift's proposal are not to be taken seriously but the enumeration of expedients with whose repetition he has no patience should be. These are roughly the same sorts of responses to the growth of populations that have been made repeatedly for the past century or more. The population problem, I must repeat, is *the* problem that requires solution before the outlines of a multitude of other problems will even come into focus.

The final selection is an essay by Thoreau. Its relation to man and his needs is subtle but profound. Planned obsolescence accounts for much of what man devours. We spend millions of dollars grinding out baubles and billions convincing persons that these trivia are the very necessities of life. Tremendous quantities of energy, real energy and not just nervous energy, are spent — literally squandered — keeping the economic machinery in motion. We still sing "Work, for the Night is Coming" although automation has made a large fraction of the nation's labor force superfluous. The questions Thoreau raises are excellent ones even a century or more after he jotted them down. The faults he castigated have in the meantime enlarged. In the last century Thoreau objected to the needless clearing of a woodlot; today he would scream in rage and horror at the strip miner and his Bunyanesque plow. Yesterday Thoreau denied that he would run around the corner to see the world blow up; today TV coverage would dog him into the innermost retreat of his home and the transistor radio would carry the news to the very center of Walden Pond. Thoreau does not touch on population problems directly in "Life Without Principle" but, until we honestly understand what he says in this essay, we too shall not be touching directly on these problems. Industry, business, and Madison Avenue become scapegoats all too easily as they do, perhaps, in the essays that follow; Thoreau points his finger directly at those responsible for society's problems — he points at you and me.

Humanae Vitae
(Of Human Life)

Pope Paul VI

Venerable Brothers and Beloved Sons:
The Transmission of Life

1. The most serious duty of transmitting human life, for which married persons are the free and responsible collaborators of God the Creator, has always been a source of great joys to them even if sometimes accompanied by not a few difficulties and by distress.

At all times the fulfillment of this duty has posed grave problems to the conscience of married persons, but with the recent evolution of society changes have taken place that give rise to new questions which the church could not ignore, having to do with a matter which so closely touches upon the life and happiness of men.

I. New Aspects of the Problem
and Competency of the Magisterium

New formulation of the problem

2. The changes which have taken place are in fact noteworthy and of varied kind. In the first place, there is the rapid demographic development. Fear is shown by many that world population is growing more rapidly than the available resources, with growing distress to many families and developing countries, so that the temptation for authorities to counter this danger with radical measures is great. Moreover, working and lodging conditions, as well as increased exigencies both in the economic field and in that of education, often make the proper education of an elevated number of children difficult today.

A change is also seen both in the manner of considering the person of woman and her place in society, and in the value to be attributed to conjugal love in marriage, and also in the appreciation to be made of the meaning of conjugal acts in relation to that love.

Finally and above all, man has made stupendous progress in the domination and rational organization of the forces of nature, such that he tends to extend this domination to his own total being: to the body, to psychical life, to social life and even to the laws which regulate the transmission of life.

3. This new state of things gives rise to new questions. Granted the conditions of life today, and granted the meaning which conjugal relations have with respect to the harmony between husband and wife and to their mutual fidelity, would not a revision of the ethical norms in force up to now seem to be advisable, especially when it is considered that they cannot be observed without sacrifices, sometimes heroic sacrifices?

And again: by extending to this field the application of the so-called "principle of totality," could it not be admitted that the intention of a less abundant but more rationalized fecundity might transform a materially sterilizing intervention into a licit and wise control of birth? Could it not be admitted, that is, that the finality of procreation pertains to the ensemble of conjugal life, rather than to its single acts? It is also asked, whether, in view of the increased sense of responsibility of modern man, the moment has not come for him to entrust to his reason and his will, rather than to the biological rhythms of his organism, the task of regulating birth.

4. Such questions require from the teaching authority of the church a new and deeper reflection upon the principles of the moral teaching on marriage: a teaching founded on the natural law, illuminated and enriched by divine revelation.

No believer will wish to deny that the teaching authority of the church is competent to interpret even the natural moral law. It is, in fact, indisputable, as our predecessors have many times declared, that Jesus Christ, when communicating to Peter and to the Apostles His divine authority and sending them to teach all nations His Commandments, constituted them as guardians and authentic interpreters of all the moral law, not only, that is, of the law of the Gospel, but also of the natural law, which is also an expression of the will of God, the faithful fulfillment of which is equally necessary for salvation.

Conformably to this mission of hers, the church has always provided — and even more amply in recent times — a coherent teaching concerning both the nature of marriage and the correct use of conjugal rights and the duties of husband and wife.

Special studies

5. The consciousness of that same mission induced us to confirm and enlarge the study commission which our predecessor Pope John XXIII of happy

memory had instituted in March, 1963. That commission, which included, besides several experts in the various pertinent disciplines, also married couples, had as its scope the gathering of opinions on the new questions regarding conjugal life, and in particular on the regulation of births, and of furnishing opportune elements of information so that the Magisterium could give an adequate reply to the expectation not only of the faithful but also of world opinion.

The work of these experts, as well as the successive judgments and counsels spontaneously forwarded by or expressly requested from a good number of our brothers in the episcopate, have permitted us to measure more exactly all the aspects of this complex matter. Hence with all our heart we express to each of them our lively gratitude.

Reply of the Magisterium

6. The conclusions at which the commission arrived could not, nevertheless, be considered by us as definitive, nor dispense us from a personal examination of this serious question; and this also because, within the commission itself, no full concordance of judgments concerning the moral norms to be proposed had been reached, and above all because certain criteria of solutions had emerged which departed from the moral teaching on marriage proposed with constant firmness by the teaching authority of the church.

Therefore, having attentively sifted the documentation laid before us, after mature reflection and assiduous prayers, we now intend, by virtue of the mandate entrusted to us by Christ, to give our reply to these grave questions.

II. Doctrinal Principles

A total vision of man

7. The problem of birth like every other problem regarding human life, is to be considered beyond partial perspectives whether of the biological or psychological, demographic or sociological orders — in the light of an integral vision of man and of his vocation, not only his natural and earthly, but also his supernatural and eternal, vocation. And since, in the attempt to justify artificial methods of birth control, many have appealed to the demands both of conjugal love and of "responsible parenthood," it is good to state very precisely the true concept of these two great realities of married life, referring principally to what was recently set forth in this regard in a highly authoritative form, by the Second Vatican Council in its pastoral constitution "Gaudium et Spes."

Conjugal love

8. Conjugal love reveals its true nature and nobility when it is considered in its supreme origin, God, who is love, "the Father, from whom every family in heaven and on earth is named."

Marriage is not then, the effect of chance or the product of evolution or unconscious natural forces; it is the wise institution of the Creator to realize in mankind his design of love. By means of the reciprocal personal gift of self, proper and exclusive to them, husband and wife tend toward the communion of their beings in view of mutual personal perfection to collaborate with God in the generation and education of new lives.

For baptized persons, moreover, marriage invests the dignity of a sacramental sign of grace, inasmuch as it represents the union of Christ and of the church.

Its characteristics

9. Under this light, there clearly appear the characteristic marks and demands of conjugal love, and it is of supreme importance to have an exact idea of these.

This love is first of all fully human, that is to say, of the senses and of the spirit at the same time. It is not, then, a simple transport of instinct and sentiment, but also, and principally, an act of the free will, intended to endure and to grow by means of the joys and sorrows of daily life in such a way that husband and wife become one only heart and one only soul, and together attain their human perfection.

Then this love is total; that is to say, it is a very special form of personal friendship, in which husband and wife generously share everything, without undue reservations or selfish calculations. Whoever truly loves his marriage partner loves not only for what he receives, but for the partner's self, rejoicing that he can enrich his partner with the gift of himself.

Again, this love is faithful and exclusive until death. Thus in fact do bride and groom conceive it to be on the day when they freely and in full awareness assume the duty of the marriage bond.

A fidelity, this, which can sometimes be difficult, but is always possible, always noble and meritorious, as no one can deny. The example of so many married persons down through the centuries shows not only that fidelity is according to the nature of marriage but also that it is a source of profound and lasting happiness.

And finally, this love is fecund, for it is not exhausted by the communion between husband and wife, but is destined to continue, raising up new lives. "Marriage and conjugal love are by their nature ordained toward the begetting and educating of children. Children are really the supreme gift of marriage and contribute very substantially to the welfare of their parents."

Responsible parenthood

10. Hence conjugal love requires in husband and wife an awareness of their mission of "responsible parenthood," which today is rightly much insisted upon, and which also must be exactly understood. Consequently it is to be considered under different aspects which are legitimate and connected with one another.

In relation to the biological processes, responsible parenthood means the knowledge and respect of their functions; human intellect discovers in the power of giving life biological laws which are part of the human person.

In relation to the tendencies of instinct or passion, responsible parenthood means that necessary dominion which reason and will must exercise over them.

In relation to physical, economic, psychological and social conditions, responsible parenthood is exercised, either by the deliberate and generous decision to raise a numerous family, or by the decision, made for grave motives and with due respect for the moral law, to avoid for the time being, or even for an indeterminate period, a new birth.

Responsible parenthood also and above all implies a more profound relationship to the moral order established by God, of which a right conscience is the faithful interpreter. The responsible exercise of parenthood implies, therefore, that husband and wife recognize fully their own duties toward God, toward themselves, toward the family and toward society, in a correct hierarchy of values.

In the task of transmitting life, therefore, they are not free to proceed completely at will, as if they could determine in a wholly autonomous way the honest path to follow; but they must conform their activity to the creative intention of God, expressed in the very nature of marriage and of its acts, and manifested by the constant teaching of the church.

Respect for the nature
and purposes of the marriage act

11. These acts, by which husband and wife are united in chaste intimacy and by means of which human life is transmitted, are, as the council recalled, "noble and worthy," and they do not cease to be lawful if, for causes independent of the will of husband and wife, they are foreseen to be infecund, since they always remain ordained toward expressing and consolidating their union. In fact, as experience bears witness, not every conjugal act is followed by a new life. God has wisely disposed natural laws and rhythms of fecundity

which, of themselves, cause a separation in the succession of births. Nonetheless the church, calling men back to the observance of the norms of the natural law, as interpreted by her constant doctrine, teaches that each and every marriage act ("quilibet matrimonii usus") must remain open to the transmission of life.

Two inseparable aspects:
union and procreation

12. That teaching, often set forth by the Magisterium, is founded upon the inseparable connection, willed by God and unable to be broken by man on his own initiative, between the two meanings of the conjugal act: the unitive meaning and the procreative meaning. Indeed, by its intimate structure, the conjugal act, while most closely uniting husband and wife, capacitates them for the generation of new lives, according to laws inscribed in the very being of man and of woman. By safeguarding both these essential aspects, the unitive and the procreative, the conjugal act preserves in its fullness the sense of true mutual love and its ordination toward man's most high calling to parenthood. We believe that the men of our day are particularly capable of seizing the deeply reasonable and human character of this fundamental principle.

Faithfulness to God's design

13. It is in fact justly observed that a conjugal act imposed upon one's partner without regard for his or her condition and lawful desires is not a true act of love, and therefore denies an exigency of right moral order in the relationship between husband and wife. Hence, one who reflects well must also recognize that a reciprocal act of love, which jeopardizes the disponibility to transmit life which God the Creator, according to particular laws, inserted therein, is in contradiction with the design constitutive of marriage, and with the will of the author of life. To use this divine gift, destroying, even if only partially, its meaning and its purpose, is to contradict the nature both of man and of woman and of their most intimate relationship, and therefore it is to contradict also the plan of God and His will.

On the other hand, to make use of the gift of conjugal love while respecting the laws of the generative process means to acknowledge oneself not to be the arbiter of the sources of human life, but rather the minister of the design established by the Creator. In fact just as man does not have unlimited dominion over his body in general, so also, with particular reason, he has no such dominion over his creative faculties as such, because of their intrinsic ordination toward raising up life, of which God is the principle. "Human life is sacred," Pope John XXIII recalled; "from its very inception it reveals the creating hand of God."

Illicit ways of regulating birth

14. In conformity with these landmarks in the human and Christian vision of marriage, we must once again declare that the direct interruption of the generative process already begun, and, above all, directly willed and procured abortion, even if for therapeutic reasons, are to be absolutely excluded as licit means of regulating birth.

Equally to be excluded, as the teaching authority of the church has frequently declared, is direct sterilization, whether perpetual or temporary, whether of the man or of the woman.

Similarly excluded in every action which, either in anticipation of the conjugal act or in its accomplishment, or in the development of its natural consequences, proposes, whether as an end or as a means, to render procreation impossible.

To justify conjugal acts made intentionally infecund, one cannot invoke as valid reasons the lesser evil, or the fact that such acts would constitute a whole together with the fecund acts already performed or to follow later, and hence would share in one and the same moral goodness. In truth, if it is sometimes licit to tolerate a lesser evil in order to avoid a greater evil or to promote a greater good, it is not licit, even for the gravest reasons, to do evil so that good may follow therefrom; that is, to make into the object of a positive act of the will something which is intrinsically disorder and hence unworthy of the human person, even when the intention is to safeguard or promote individual, family or social well-being.

Consequently it is an error to think that a conjugal act which is deliberately made infecund and so is intrinsically dishonest could be made honest and right by the ensemble of a fecund conjugal life.

Licitness of therapeutic means

15. The church, on the contrary, does not at all consider illicit the use of those therapeutic means truly necessary to cure diseases of the organism, even if an impediment to procreation, which may be foreseen, should result therefrom, provided such impediment is not, for whatever motive, directly willed.

Licitness of recourse
to infecund periods

16. To this teaching of the church on conjugal morals, the objection is made today, as we observed earlier, that it is the prerogative of the human intellect to dominate the energies offered by irrational nature and to orientate them toward an end conformable to the good of man. Now, some may ask: In

the present case, is it not reasonable in many circumstances to have recourse to artificial birth control if, thereby, we secure the harmony and peace of the family, and better conditions for the education of the children already born? To this question it is necessary to reply with clarity: The church is the first to praise and recommend the intervention of intelligence in a function which so closely associates the rational creature with his Creator, but she affirms that this must be one with respect for the order established by God.

If, then, there are serious motives to space out births, which derive from the physical or psychological conditions of husband and wife, or from external conditions, the church teaches that it is then licit to take into account the natural rhythms immanent in the generative functions, for the use of marriage in the infecund periods only, and this way to regulate birth without offending the moral principles which have been recalled earlier.

The church is coherent with herself when she considers recourse to the infecund periods to be licit, while at the same time condemning, as being always illicit, the use of means directly contrary to fecundation, even if such use is inspired by reasons which may appear honest and serious. In reality, there are essential differences between the two cases: In the former, the married couple make legitimate use of a natural disposition; in the latter, they impede the development of natural processes. It is true that, in the one and the other case, the married couple are concordant in the positive will of avoiding children for plausible reasons, seeking the certainty that offspring will not arrive; but it is also true that only in the former case are they able to renounce the use of marriage in the fecund periods when, for just motives, procreation is not desirable, while making use of it during infecund periods to manifest their affection and to safeguard their mutual fidelity. By so doing, they give proof of a truly and integrally honest love.

Grave consequences of
methods of artificial birth control

17. Upright men can even better convince themselves of the solid grounds on which the teaching of the church in this field is based, if they care to reflect upon the consequences of methods of artificial birth control. Let them consider, first of all, how wide and easy a road would thus be opened up toward conjugal infidelity and the general lowering of morality. Not much experience is needed in order to know human weakness, and to understand that men — especially the young, who are so vulnerable on this point — have need of encouragement to be faithful to the moral law, so that they must not be offered some easy means of eluding its observance. It is also to be feared that the man, growing used to the employment of anticonceptive practices, may finally lose respect for the woman and, no longer caring for her physical and psychological equilibrium, may come

to the point of considering her as a mere instrument of selfish enjoyment, and no longer as his respected and beloved companion.

Let it be considered also that a dangerous weapon would thus be placed in the hands of those public authorities who take no heed of moral exigencies. Who could blame a government for applying to the solution of the problems of the community those means acknowledged to be licit for married couples in the solution of a family problem? Who will stop rulers from favoring, from even imposing upon their peoples, if they were to consider it necessary, the method of contraception which they judge to be most efficacious? In such a way men, wishing to avoid individual, family or social difficulties encountered in the observance of the divine law, would reach the point of placing at the mercy of the intervention of public authorities the most personal and most reserved sector of conjugal intimacy.

Consequently, if the mission of generating life is not to be exposed to the arbitrary will of men, one must necessarily recognize insurmountable limits to the possibility of man's domination over his own body and its functions; limits which no man, whether a private individual or one invested with authority, may licitly surpass. And such limits cannot be determined otherwise than by the respect due to the integrity of the human organism and its functions, according to the principles recalled earlier, and also according to the correct understanding of the "principle of totality" illustrated by our predecessor Pope Pius XII.

The church guarantor
of true human values

18. It can be foreseen that this teaching will perhaps not be easily received by all: too numerous are those voices — amplified by the modern means of propaganda — which are contrary to the voice of the church. To tell the truth, the church is not surprised to be made, like her divine Founder, a "sign of contradiction"; yet she does not because of this cease to proclaim with humble firmness the entire moral law, both natural and evangelical. Of such laws the church was not the author, nor consequently can she be their arbiter; she is only their depositary and their interpreter, without ever being able to declare to be licit that which is not so by reasons of its intimate and unchangeable opposition to the true good of man.

In defending conjugal morals in their integral wholeness, the church knows that she contributes towards the establishment of a truly human civilization; she engages man not to abdicate from his own responsibility in order to rely on technical means; by that very fact she defends the dignity of man and wife. Faithful to both the teaching and the example of the Saviour, she shows herself to be the sincere and disinterested friend of men, whom she wishes to help, even during their earthly sojourn, "to share as sons in the life of living God, the Father of all men."

III. Pastoral Directives

The church mater et magistra

19. Our words would not be an adequate expression of the thought and solicitude of the church, mother and teacher of all peoples, if, after having recalled men to the observance and respect of the divine law regarding matrimony, we did not strengthen them in the path of honest regulation of birth, even amid the difficult conditions which today afflict families and peoples.

The church in fact, cannot have a different conduct toward men than that of the redeemer: she knows their weaknesses, has compassion on the crowd, receives sinners; but she cannot renounce the teaching of the law which is, in reality, that law proper to a human life restored to its original truth and conducted by the spirit of God.

Though we are thinking also of all men of goodwill, we now address ourself particularly to our sons, from whom we expect a prompter and more generous adherence.

Possibility of observing the divine law

20. The teaching of the church on the regulation of birth, which promulgates the divine law, will easily appear to many to be difficult or even impossible of actuation. And indeed, like all great beneficent realities, it demands serious engagement and much effort, individual, family and social effort. More than that, it would not be practicable without the help of God, who upholds and strengthens the goodwill of men. Yet, to anyone who reflects well, it cannot but be clear that such efforts ennoble man and are beneficial to the human community.

Mastery of self

21. The honest practice of regulation of birth demands first of all that husband and wife acquire and possess solid convictions concerning the true

values of life of the family, and that they tend toward securing perfect self-mastery. To dominate instinct by means of one's reason and free will undoubtedly requires ascetical practices, so that the affective manifestations of conjugal life may observe the correct order, in particular with regard to the observance of periodic continence.

Yet this discipline which is proper to the purity of married couples, far from harming conjugal love, rather confers on it a higher human value. It demands continual effort yet, thanks to its beneficent influence, husband and wife fully develop their personalities, being enriched with spiritual values. Such discipline bestows upon family life fruits of serenity and peace, and facilitates the solution of other problems; it favors attention for one's partner, helps both parties to drive out selfishness, the enemy of true love; and deepens their sense of responsibility. By its means, parents acquire the capacity of having a deeper and more efficacious influence in the education of their offspring; little children and youths grow up with a just appraisal of human values, and in the serene and harmonious development of their spiritual and sensitive faculties.

Creating an atmosphere
favorable to chastity

22. On this occasion, we wish to draw the attention of educators, and of all who perform duties of responsibility in regard to the common good of human society, to the need of creating an atmosphere favorable to education in chastity, that is, to the triumph of healthy liberty over license by means of respect for the moral order.

Everything in the modern media of social communications which leads to sensed excitation and unbridled customs, as well as every form of pornography and licentious performances, must arouse the frank and unanimous reaction of all those who are solicitous for the progress of civilization and the defense of the supreme good of the human spirit. Vainly would one seek to justify such depravation with the pretext of artistic or scientific exigencies, or to deduce an argument from the freedom allowed in this sector by the public authorities.

Appeal to public authorities

23. To rulers, who are those principally responsible for the common good, and who can do so much to safeguard moral customs, we say: Do not allow the morality of your peoples to be degraded; do not permit that by legal means practices contrary to the natural and divine law be introduced into that fundamental cell, the family. Quite other is the way in which public authorities can and must contribute to the solution of the demographic problem: namely,

the way of a provident policy for the family, of a wise education of peoples in respect of the moral law and the liberty of citizens.

We are well aware of the serious difficulties experienced by public authorities in this regard, especially in the developing countries. To their legitimate preoccupations we devoted our encyclical letter "Populorum Progressio." But, with our predecessor Pope John XXIII, we repeat: No solution to these difficulties is acceptable "which does violence to man's essential dignity" and is based only "on an utterly materialistic conception of man himself and of his life." The only possible solution to this question is one which envisages the social and economic progress both of individuals and of the whole of human society, and which respects and promotes true human values.

Neither can one, without grave injustice, consider Divine Providence to be responsible for what depends, instead, on a lack of wisdom in government, on an insufficient sense of social justice, on selfish monopolization or again on blameworthy indolence in confronting the efforts and the sacrifices necessary to insure the raising of living standards of a people and of all its sons.

May all responsible public authorities — as some are already doing so laudably — generously revive their efforts. And may mutual aid between all the members of the great human family never cease to grow: This is an almost limitless field which thus opens up to the activity of the great international organizations.

To men of science

24. We wish to express our encouragement to men of science, who "can considerably advance the welfare of marriage and the family, along with peace of conscience, if by pooling their efforts they labor to explain more thoroughly the various conditions favoring a proper regulation of births." It is particularly desirable that, according to the wish already expressed by Pope Pius XII, medical science succeed in providing a sufficiently secure basis for a regulation of birth, founded on the observance of natural rhythms. In this way, scientists and especially Catholic scientists will contribute to demonstrate in actual fact that, as the church teaches, "a true contradiction cannot exist between the divine laws pertaining to the transmission of life and those pertaining to the fostering of authentic conjugal love."

To Christian husbands and wives

25. And now our words more directly address our own children, particularly those whom God calls to serve him in marriage. The church, while teaching imprescriptible demands of the divine law, announces the tidings of

salvation, and by means of the sacraments opens up the paths of grace, which makes man a new creature, capable of corresponding with love and true freedom to the design of his creator and Saviour, and of finding the yoke of Christ to be sweet.

Christian married couples, then, docile to her voice, must remember that their Christian vocation, which began at baptism, is further specified and reinforced by the sacrament of matrimony.

By it husband and wife are strengthened and, as it were, consecrated for the faithful accomplishment of their proper duties, for the carrying out of their proper vocation even to perfection, and the Christian witness which is proper to them before the whole world. To them the Lord entrusts the task of making visible to men the holiness and sweetness of the law which unites the mutual love of husband and wife with their cooperation with the love of God the author of human life.

We do not at all intend to hide the sometimes serious difficulties inherent in the life of Christian married persons; for them as for everyone else, "the gate is narrow and the way is hard, that leads to life." But the hope of that life must illuminate their way, as with courage they strive to live with wisdom, justice and piety in this present time, knowing that the figure of this world passes away.

Let married couples, then, face up to the efforts needed, supported by the faith and hope which "do not disappoint. . . because God's love has been poured into our hearts through the Holy Spirit, who has been given to us"; let them implore divine assistance by persevering prayer; above all, let them draw from the source of grace and charity in the eucharist. And if sin should still keep its hold over them, let them not be discouraged, but rather have recourse with humble perseverance to the mercy of God, which is poured forth in the sacrament of penance. In this way they will be enabled to achieve the fullness of conjugal life described by the Apostle: "Husbands, love your wives, as Christ loved the church. . . .Husbands should love their wives as their own bodies. He who loves his wife loves himself. For no man ever hates his own flesh, but nourishes and cherishes it, as Christ does the church. . . .This is a great mystery, and I mean in reference to Christ and the church. However, let each one of you love his wife as himself, and let the wife see that she respects her husband."

Apostolate in homes

26. Among the fruits which ripen forth from a generous effort of fidelity to the divine law, one of the most precious is that married couples themselves not infrequently feel the desire to communicate their experience to others. Thus there comes to be included in the vast pattern of the vocation of the laity a new and most noteworthy form of the apostolate of like to like: It is married couples themselves who become apostles and guides to other married couples. This is

assuredly, among so many forms of apostolate, one of those which seem most opportune today.

To doctors and medical personnel

27. We hold those physicians and medical personnel in the highest esteem who, in the exercise of their profession, value above every human interest the superior demands of their Christian vocation. Let them persevere, therefore, in promoting on every occasion the discovery of solutions inspired by faith and right reason, let them strive to arouse this conviction and this respect in their associates. Let them also consider as their proper professional duty the task of acquiring all the knowledge needed in this delicate sector, so as to be able to give to those married persons who consult them wise counsel and healthy direction, such as they have a right to expect.

To priests

28. Beloved priest sons, by vocation you are the counselors and spiritual guides of individual persons and of families. We now turn to you with confidence. Your first task — especially in the case of those who reach moral theology — is to expound the church's teaching on marriage without ambiguity. Be the first to give, in the exercise of your ministry, the example of loyal internal and external obedience to the teaching authority of the church. That obedience, as you know well, obliges not only because of the reasons adduced, but rather because of the light of the Holy Spirit, which is given in a particular way to the pastors of the church in order that they may illustrate the truth. You know, too, that it is of the utmost importance, for peace of consciences and for the unity of the Christian people, that in the field of morals as well as in that of dogma, all should attend to the Magisterium of the church, and all should speak the same language. Hence, with all our heart we renew to you the heartfelt plea of the great Apostle Paul: "I appeal to you, brethren, by the name of our Lord Jesus Christ, that all of you agree and that there be no dissensions among you, but that you be united in the same mind and the same judgment."

29. To diminish in no way the saving teaching of Christ constitutes an eminent form of charity for souls. But this must ever be accompanied by patience and goodness, such as the Lord Himself gave example of in dealing with men. Having come not to condemn but to save, he was indeed intransigent with evil but merciful toward individuals.

In their difficulties, may married couples always find, in the words and in the heart of a priest, the echo of the voice and the love of the Redeemer.

To bishops

30. Beloved and venerable brothers in the episcopate, with whom we most intimately share the solicitude of the spiritual good of the people of God, at the conclusion of this encyclical our reverent and affectionate thoughts turn to you. To all of you we extend an urgent invitation. At the head of the priests, your collaborators, and of your faithful, work ardently and incessantly for the safeguarding and the holiness of marriage, so that it may always be lived in its entire human and Christian fullness. Consider this mission as one of your most urgent responsibilities at the present time. As you know, it implies concerted pastoral action in all the fields of human activity, economic, cultural and social: for, in fact, only a simultaneous improvement in these various sectors will make it possible to render the life of parents and of children within their families not only tolerable, but easier and more joyous, to render the living together in human society more fraternal and peaceful, in faithfulness to God's design for the world.

Final appeal

Venerable brothers, most beloved sons and all men of goodwill, great indeed is the work of education, of progress and of love to which we call you, upon the foundation of the church's teaching, of which the successor of Peter is, together with his brothers in the episcopate, the depositary and interpreter. Truly a great work, as we are deeply convinced. Both for the world and for the church, since man cannot find true happiness — toward which he aspires with all his being — other than in respect of the laws written by God in His very nature, laws which he must observe with intelligence and love. Upon this work, and upon all of you, and especially upon married couples, we invoke the abundant graces of the God of holiness and mercy, and in pledge thereof we impart to you all our apostolic blessing.

A Modest Proposal

FOR PREVENTING THE CHILDREN OF POOR PEOPLE IN
IRELAND FROM BEING A BURDEN TO THEIR
PARENTS OR COUNTRY, AND FOR MAKING THEM
BENEFICIAL TO THE PUBLIC

Jonathan Swift

It is a melancholy object to those who walk through this great town, or travel in the country, when they see the streets, the roads and cabin-doors crowded with beggars of the female sex, followed by three, four, or six children, all in rags, and importuning every passenger for an alms. These mothers, instead of being able to work for their honest livelihood, are forced to employ all their time in strolling, to beg sustenance for their helpless infants, who, as they grow up, either turn thieves for want of work, or leave their dear native country to fight for the Pretender in Spain, or sell themselves to the Barbadoes.

I think it is agreed by all parties that this prodigious number of children, in the arms, or on the backs, or at the heels of their mothers, and frequently of their fathers, is in the present deplorable state of the kingdom a very great additional grievance; and therefore whoever could find out a fair, cheap, and easy method of making these children sound and useful members of the commonwealth would deserve so well of the public as to have his statue set up for a preserver of the nation.

But my intention is very far from being confined to provide only for the children of professed beggars; it is of a much greater extent, and shall take in the whole number of infants at a certain age who are born of parents in effect as little able to support them as those who demand our charity in the streets.

As to my own part, having turned my thoughts for many years upon this important subject, and maturely weighed the several schemes of other projectors, I have always found them grossly mistaken in their computation. It is true a child just dropped from its dam may be supported by her milk for a solar year with little other nourishment, at most not above the value of two shillings, which the mother may certainly get, or the value in scraps, by her lawful occupation of begging, and it is exactly at one year old that I propose to provide for them, in such a manner as, instead of being a charge upon their parents, or the parish, or wanting food and raiment for the rest of their lives, they shall, on the contrary, contribute to the feeding and partly to the clothing of many thousands.

There is likewise another great advantage in my scheme, that it will prevent those voluntary abortions, and that horrid practice of women murdering their bastard children, alas, too frequent among us, sacrificing the poor innocent babes, I doubt, more to avoid the expense than the shame, which would move tears and pity in the most savage and inhuman breast.

The number of souls in Ireland being usually reckoned one million and a half, of these I calculate there may be about two hundred thousand couples whose wives are breeders, from which number I subtract thirty thousand couples who are able to maintain their own children, although I apprehend there cannot be so many under the present distresses of the kingdom, but this being granted, there will remain an hundred and seventy thousand breeders. I again subtract fifty thousand for those women who miscarry, or whose children die by accident or disease within the year. There only remain an hundred and twenty thousand children of poor parents annually born: the question therefore is, how this number shall be reared, and provided for, which, as I have already said, under the present situation of affairs is utterly impossible by all the methods hitherto proposed, for we can neither employ them in handicraft or agriculture; we neither build houses (I mean in the country), nor cultivate land: they can very seldom pick up a livelihood by stealing until they arrive at six years old, except where they are of towardly parts, although I confess they learn the rudiments much earlier, during which time they can however be properly looked upon only as probationers, as I have been informed by a principal gentleman in the County of Cavan, who protested to me that he never knew above one or two instances under the age of six, even in a part of the kingdom so renowned for the quickest proficiency in that art.

I am assured by our merchants that a boy or a girl before twelve years old, is no saleable commodity, and even when they come to this age, they will not yield above three pounds, or three pounds and half-a-crown at most on the Exchange, which cannot turn to account either to the parents or the kingdom, the charge of nutriment and rags having been at least four times that value.

I shall now therefore humbly propose my own thoughts, which I hope will not be liable to the least objection.

I have been assured by a very knowing American of my acquaintance in London, that a young healthy child well nursed is at a year old a most delicious, nourishing and wholesome food, whether stewed, roasted, baked, or boiled, and I make no doubt that it will equally serve in a fricassee, or a ragout.

I do therefore humbly offer it to public consideration, that of the hundred and twenty thousand children already computed, twenty thousand may be reserved for breed, whereof only one fourth part to be males, which is more than we allow to sheep, black-cattle, or swine, and my reason is that these children are seldom the fruits of marriage, a circumstance not much regarded by our savages, therefore one male will be sufficient to serve four females. That the remaining hundred thousand may at a year old be offered in sale to the persons of quality, and fortune, through the kingdom, always advising the mother to let them suck plentifully in the last month, so as to render them plump, and fat for a good table. A child will make two dishes at an entertainment for friends, and when the family dines alone, the fore or hind quarter will make a reasonable dish, and seasoned with a little pepper or salt will be very good boiled on the fourth day, especially in winter.

I have reckoned upon a medium, that a child just born will weigh twelve pounds, and in a solar year if tolerably nursed increaseth to twenty-eight pounds.

I grant this food will be somewhat dear, and therefore very proper for landlords, who, as they have already devoured most of the parents, seem to have the best title to the children.

Infant's flesh will be in season throughout the year, but more plentiful in March, and a little before and after, for we are told by a grave* author, an eminent French physician, that fish being a prolific diet, there are more children born in Roman Catholic countries about nine months after Lent than at any other season; therefore reckoning a year after Lent, the markets will be more glutted than usual, because the number of Popish infants is at least three to one in this kingdom, and therefore it will have one other collateral advantage by lessening the number of Papists among us.

I have already computed the charge of nursing a beggar's child (in which list I reckon all cottagers, labourers, and four-fifths of the farmers) to be about two shillings per annum, rags included, and I believe no gentleman would repine to give ten shillings for the carcass of a good fat child, which, as I have said, will make four dishes of excellent nutritive meat, when he hath only some particular friend or his own family to dine with him. Thus the Squire will learn to be a good landlord and grow popular among his tenants, the mother will have eight shillings net profit, and be fit for work until she produces another child.

Those who are more thrifty (as I must confess the times require) may flay the carcass; the skin of which artifically dressed, will make admirable gloves for ladies, and summer boots for fine gentlemen.

As to our city of Dublin, shambles may be appointed for this purpose, in the most convenient parts of it, and butchers we may be assured will not be wanting, although I rather recommend buying the children alive, and dressing them hot from the knife, as we do roasting pigs.

A very worthy person, a true lover of his country, and whose virtues I highly esteem, was lately pleased, in discoursing on this matter to offer a refinement upon my scheme. He said that many gentlemen of this kingdom,

---

*Rabelais.

having of late destroyed their deer, he conceived that the want of venison might be well supplied by the bodies of young lads and maidens, not exceeding fourteen years of age, nor under twelve, so great a number of both sexes in every county being now ready to starve, for want of work and service: and these to be disposed of by their parents if alive, or otherwise by their nearest relations. But with due deference to so excellent a friend, and so deserving a patriot, I cannot be altogether in his sentiments. For as to the males, my American acquaintance assured me from frequent experience that their flesh was generally tough and lean, like that of our schoolboys, by continual exercise, and their taste disagreeable, and to fatten them would not answer the charge. Then as to the females, it would, I think with humble submission, be a loss to the public, because they soon would become breeders themselves: and besides, it is not improbable that some scrupulous people might be apt to censure such a practice (although indeed very unjustly) as a little bordering upon cruelty, which I confess, hath always been with me the strongest objection against any project, howsoever well intended.

But in order to justify my friend, he confessed that this expedient was put into his head by the famous Psalmanazar, a native of the island Formosa, who came from thence to London, above twenty years ago, and in conversation told my friend that in his country when any young person happened to be put to death, the executioner sold the carcass to persons of quality, as a prime dainty, and that, in his time, the body of a plump girl of fifteen, who was crucified for an attempt to poison the emperor, was sold to his Imperial Majesty's Prime Minister of State, and other great Mandarins of the Court, in joints from the gibbet, at four hundred crowns. Neither indeed can I deny that if the same use were made of several plump young girls in this town who, without one single groat to their fortunes, cannot stir abroad without a chair, and appear at the playhouse and assemblies in foreign fineries, which they never will pay for, the kingdom would not be the worse.

Some persons of a desponding spirit are in great concern about that vast number of poor people, who are aged, diseased, or maimed, and I have been desired to employ my thoughts what course may be taken to ease the nation of so grievous an encumbrance. But I am not in the least pain upon that matter, because it is very well known that they are every day dying, and rotting, by cold, and famine, and filth, and vermin, as fast as can be reasonably expected. And as to the younger labourers they are now in almost as hopeful a condition. They cannot get work, and consequently pine away from want of nourishment, to a degree that if at any time they are accidentally hired to common labour, they have not strength to perform it; and thus the country and themselves are in a fair way of being soon delivered from the evils to come.

I have too long digressed, and therefore shall return to my subject. I think the advantages by the proposal which I have made are obvious and many, as well as of the highest importance.

For first, as I have already observed, it would greatly lessen the number of

Papists, with whom we are yearly over-run, being the principal breeders of the nation, as well as our most dangerous enemies, and who stay at home on purpose with a design to deliver the kingdom to the Pretender, hoping to take their advantage by the absence of so many good Protestants, who have chosen rather to leave their country than stay at home and pay tithes against their conscience to an idolatrous Episcopal curate.

Secondly, the poorer tenants will have something valuable of their own, which by law may be made liable to distress, and help to pay their landlord's rent, their corn and cattle being already seized, and money a thing unknown.

Thirdly, whereas the maintenance of an hundred thousand children, from two years old, and upwards, cannot be computed at less than ten shillings a piece per annum, the nation's stock will be thereby increased fifty thousand pounds per annum, besides the profit of a new dish, introduced to the tables of all gentlemen of fortune in the kingdom, who have any refinement in taste, and the money will circulate among ourselves, the goods being entirely of our own growth and manufacture.

Fourthly, the constant breeders, besides the gain of eight shillings sterling per annum, by the sale of their children, will be rid of the charge of maintaining them after the first year.

Fifthly, this food would likewise bring great custom to taverns, where the vintners will certainly be so prudent as to procure the best receipts for dressing it to perfection, and consequently have their houses frequented by all the fine gentlemen, who justly value themselves upon their knowledge in good eating; and a skillful cook, who understands how to oblige his guests, will contrive to make it as expensive as they please.

Sixthly, this would be a great inducement to marriage, which all wise nations have either encouraged by rewards, or enforced by laws and penalties. It would increase the care and tenderness of mothers towards their children, when they were sure of a settlement for life, to the poor babes, provided in some sort by the public to their annual profit instead of expense. We should soon see an honest emulation among the married women, which of them could bring the fattest child to the market. Men would become as fond of their wives, during the time of their pregnancy, as they are now of their mares in foal, their cows in calf, or sows when they are ready to farrow, nor offer to beat or kick them (as it is too frequent a practice) for fear of a miscarriage.

Many other advantages might be enumerated. For instance, the addition of some thousand carcasses in our exportation of barrelled beef; the propagation of swine's flesh, and improvement in the art of making good bacon, so much wanted among us by the great destruction of pigs, too frequent at out tables, are no way comparable in taste or magnificence to a well-grown, fat yearling child, which roasted whole will make a considerable figure at a Lord Mayor's feast, or any other public entertainment. But this and many others I omit, being studious of brevity.

Supposing that one thousand families in this city would be constant customers for infants flesh, besides others who might have it at merry meetings, particularly weddings and christenings; I compute that Dublin would take off annually about twenty thousand carcasses, and the rest of the kingdom (where probably they will be sold somewhat cheaper) the remaining eighty thousand.

I can think of no one objection that will possibly be raised against this proposal, unless it should be urged that the number of people will be thereby much lessened in the kingdom. This I freely own, and it was indeed one principal design in offering it to the world. I desire the reader will observe, that I calculate my remedy for this one individual Kingdom of Ireland, and for no other that ever was, is, or, I think, ever can be upon earth. Therefore let no man talk to me of other expedients: Of taxing our absentees at five shillings a pound: Of using neither clothes, nor household furniture, except what is of our own growth and manufacture: Of utterly rejecting the materials and instruments that promote foreign luxury: Of curing the expensiveness of pride, vanity, idleness, and gaming in our women: Of introducing a vein of parsimony, prudence, and temperance: Of learning to love our country, wherein we differ even from Laplanders, and the inhabitants of Topinamboo: Of quitting our animosities and factions, nor act any longer like the Jews, who were murdering one another at the very moment their city was taken: Of being a little cautious not to sell our country and consciences for nothing: Of teaching landlords to have at least one degree of mercy towards their tenants. Lastly, of putting a spirit of honesty, industry, and skill into our shopkeepers, who, if a resolution could now be taken to buy only our native goods, would immediately unite to cheat and exact upon us in the price, the measure and the goodness, nor could ever yet be brought to make one fair proposal of just dealing, though often and earnestly invited to it.

Therefore I repeat, let no man talk to me of these and the like expedients, till he hath at least a glimpse of hope that there will ever be some hearty and sincere attempt to put them in practice.

But as to myself, having been wearied out for many years with offering vain, idle, visionary thoughts, and at length utterly despairing of success, I fortunately fell upon this proposal, which as it is wholly new, so it hath something solid and real, of no expense and little trouble, full in our own power, and whereby we can incur no danger in disobliging England. For this kind of commodity will not bear exportation, the flesh being too tender a consistence to admit a long continuance in salt, although perhaps I could name a country which would be glad to eat up our whole nation without it.

After all I am not so violently bent upon my own opinion as to reject any offer, proposed by wise men, which shall be found equally innocent, cheap, easy and effectual. But before some thing of that kind shall be advanced in contradiction to my scheme, and offering a better, I desire the author, or authors, will be pleased maturely to consider two points. First, as things now stand, how they will be able to find food and raiment for a hundred thousand useless mouths and backs? And secondly, there being a round million of creatures in human figure, throughout this kingdom, whose whole subsistence put into a common stock would leave them in debt two millions of pounds sterling; adding those who are beggars by profession, to the bulk of farmers, cottagers, and labourers with their wives and children, who are beggars in effect; I desire those politicians who dislike my overture, and may perhaps be so bold to attempt an answer, that they will first ask the parents of these mortals whether they would not at this day think it a great happiness to have been sold for food at a year old, in the manner I prescribe, and thereby have avoided such a perpetual scene of misfortunes as they have since gone through, by the oppression of landlords, the impossibility of paying rent without money or trade, the want of common sustenance, with neither house nor clothes to cover them from the inclemencies of weather, and the most inevitable prospect of entailing the like, or greater miseries upon their breed for ever.

I profess in the sincerity of my heart that I have not the least personal interest in endeavouring to promote this necessary work, having no other motive than the public good of my country, by advancing our trade, providing for infants, relieving the poor, and giving some pleasure to the rich. I have no children by which I can propose to get a single penny; the youngest being nine years old, and my wife past child-bearing.

Life Without Principle

Henry Thoreau

At a lyceum, not long since, I felt that the lecturer had chosen a theme too foreign to himself, and so failed to interest me as much as he might have done. He described things not in or near to his heart, but toward his extremities and superficies. There was, in this sense, no truly central or centralizing thought in the lecture. I would have had him deal with privatest experience, as the poet does. The greatest compliment that was ever paid me was when one asked me what I THOUGHT, and attended to my answer. I am surprised, as well as delighted, when this happens, it is such a rare use he would make of me, as if he were acquainted with the tool. Commonly, if men want anything of me, it is only to know how many acres I make of their land — since I am a surveyor — or, at most, what trivial news I have burdened myself with. They never will go to law for my meat; they prefer the shell. A man once came a considerable distance to ask me to lecture on Slavery; but on conversing with him, I found that he and his clique expected seven eighths of the lecture to be theirs, and only one eighth mine; so I declined. I take it for granted, when I am invited to lecture anywhere — for I have had a little experience in that business — that there is a desire to hear what I THINK on some subject, though I may be the greatest fool in the country, and not that I should say pleasant things merely, or such as the audience will assent to; and I resolve, accordingly, that I will give them a strong dose of myself. They have sent for me, and engaged to pay for me, and I am determined that they shall have me, though I bore them beyond all precedent.

So now I would say something similar to you, my readers. Since you are my readers, and I have not been much of a traveler, I will not talk about people a thousand miles off, but come as near home as I can. As the time is short, I will leave out all the flattery, and retain all the criticism.

Let us consider the way in which we spend our lives.

This world is a place of business. What an infinite bustle! I am awaked almost every night by the panting of the locomotive. It interrupts my dreams. There is no sabbath. It would be glorious to see mankind at leisure for once. It is nothing but work, work, work. I cannot easily buy a blankbook to write thoughts in; they are commonly ruled for dollars and cents. An Irishman, seeing me making a minute in the fields, took it for granted that I was calculating my wages. If a man was tossed out of a window when an infant, and so made a cripple for life, or scared out of his wits by the Indians, it is regretted chiefly because he was thus incapacitated for — business! I think that there is nothing, not even crime, more opposed to poetry, to philosophy, ay, to life itself, than this incessant business.

There is a coarse and boisterous money-making fellow in the outskirts of our town, who is going to build a bank-wall under the hill along the edge of his meadow. The powers have put this into his head to keep him out of mischief, and he wishes me to spend three weeks digging there with him. The result will be that he will perhaps get some more money to hoard, and leave for his heirs to spend foolishly. If I do this, most will commend me as an industrious and hard-working man; but if I choose to devote myself to certain labors which yield more real profit, though but little money, they may be inclined to look on me as an idler. Nevertheless, as I do not need the police of meaningless labor to regulate me, and do not see anything absolutely praiseworthy in this fellow's undertaking any more than in many an enterprise of our own or foreign governments, however amusing it may be to him or them, I prefer to finish my education at a different school.

If a man walk in the woods for love of them half of each day, he is in danger of being regarded as a loafer; but if he spends his whole day as a speculator, shearing off those woods and making earth bald before her time, he is esteemed an industrious and enterprising citizen. As if a town had no interest in its forests but to cut them down!

Most men would feel insulted if it were proposed to employ them in throwing stones over a wall, and then in throwing them back, merely that they might earn their wages. But many are no more worthily employed now. For instance: just after sunrise, one summer morning, I noticed one of my neighbors walking beside his team, which was slowly drawing a heavy hewn stone swung under the axle, surrounded by an atmosphere of industry — his day's work begun, his brow commenced to sweat — a reproach to all sluggards and idlers — pausing abreast the shoulders of his oxen, and half turning round with a flourish of his merciful whip, while they gained their length of him. And I thought, Such is the labor which the American Congress exists to protect — honest, manly toil — honest as the day is long — that makes his bread taste sweet, and keeps society sweet — which all men respect and have consecrated; one of the sacred band, doing the needful but irksome drudgery. Indeed, I felt a slight reproach, because I observed this from a window, and was not abroad and stirring about a similar business. The day went by, and at evening I passed the yard of another neighbor, who keeps many servants, and spends much money foolishly, while he adds nothing to the common stock, and there I saw the stone of the morning lying beside a whimsical structure intended to adorn this Lord Timothy Dexter's premises, and the dignity forthwith departed from the teamster's labor, in my eyes. In my opinion, the sun was made to light worthier toil than this. I may add that his employer has since run off, in debt to a good part of the town, and, after passing through Chancery, has settled somewhere else, there to become once more a patron of the arts.

The ways by which you may get money almost without exception lead downward. To have done anything by which you earned money MERELY is to have been truly idle or worse. If the laborer gets no more than the wages which his employer pays him, he is cheated, he cheats himself. If you would get money as a writer or lecturer, you must be popular, which is to go down perpendicularly. Those services which the community will most readily pay for, it is most disagreeable to render. You are paid for being something less than a man. The state does not commonly reward a genius any more wisely. Even the poet laureate would rather not have to celebrate the accidents of royalty. He must be bribed with a pipe of wine; and perhaps another poet is called away from his muse to gauge that very pipe. As for my own business, even that kind of surveying which I could do with most satisfaction my employers do not want. They would prefer that I should do my work coarsely and not too well, ay, not well enough. When I observe that there are different ways of surveying, my employer commonly asks which will give him the most land, not which is most correct. I once invented a rule for measuring cordwood, and tried to introduce it in Boston; but the measurer there told me that the sellers did not wish to have their wood measured correctly — that he was already too accurate for them, and therefore they commonly got their wood measured in Charlestown before crossing the bridge.

The aim of the laborer should be, not to get his living, to get "a good job," but to perform well a certain work; and, even in a pecuniary sense, it would be economy for a town to pay its laborers so well that they would not feel that they were working for low ends, as for a livelihood merely, but for scientific, or even moral ends. Do not hire a man who does your work for money, but him who does it for love of it.

It is remarkable that there are few men so well employed, so much to their minds, but that a little money or fame would commonly buy them off from their present pursuit. I see advertisements for active young men, as if activity were the whole of a young man's capital. Yet I have been surprised when one has with confidence proposed to me, a grown man, to embark in some enterprise of his, as if I had absolutely nothing to do, my life having been a complete failure hitherto. What a doubtful compliment this to pay me! As if he had met me halfway across the ocean beating up against the wind, but bound nowhere, and proposed to me to go along with him! If I did, what do you think the underwriters would say? No, no! I am not without employment at this stage of the voyage. To tell the truth, I saw an advertisement for able-bodied seamen, when I was a boy, sauntering in my native port, and as soon as I came of age I embarked.

The community has no bribe that will tempt a wise man. You may raise money enough to tunnel a mountain, but you cannot raise money enough to hire a man who is minding his own business. An efficient and valuable man does what he can, whether the community pay him for it or not. The inefficient offer their inefficiency to the highest bidder, and are forever expecting to be put into office. One would suppose that they were rarely disappointed.

Perhaps I am more than usually jealous with respect to my freedom. I feel that my connection with and obligation to society are still very slight and transient. Those slight labors which afford me a livelihood, and by which it is allowed that I am to some extent serviceable to my contemporaries, are as yet commonly a pleasure to me, and I am not often reminded that they are a necessity. So far I am successful. But I foresee that if my wants should be much increased, the labor required to supply them would become a drudgery. If I should sell both my forenoons and afternoons to society, as most appear to do, I am sure that for me there would be nothing left worth living for. I trust that I shall never thus sell my birthright for a mess of pottage. I wish to suggest that a man may be very industrious, and yet not spend his time well. There is no more fatal blunderer than he who consumes the greater part of his life getting his living. All great enterprises are self-supporting. The poet, for instance, must sustain his body by his poetry, as a steam planing-mill feeds its boilers with the shavings it makes. You must get your living by loving. But as it is said of the merchants that ninety-seven in a hundred fail, so the life of men generally, tried by this standard, is a failure, and bankruptcy may be surely prophesied.

Merely to come into the world the heir of a fortune is not to be born, but to be stillborn, rather. To be supported by the charity of friends, or a government pension — provided you continue to breathe — by whatever fine synonyms you describe these relations, is to go into the almshouse. On Sundays the poor debtor goes to church to take an account of stock, and finds, of course, that his outgoes have been greater than his income. In the Catholic Church, especially, they go into chancery, make a clean confession, give up all, and think to start again. Thus men will lie on their backs, talking about the fall of man, and never make an effort to get up.

As for the comparative demand which men make on life, it is an important difference between two, that the one is satisfied with a level success, that his marks can all be hit by point-blank shots, but the other, however low and unsuccessful his life may be, constantly elevates his aim, though at a very slight angle to the horizon. I should much rather be the last man, though, as the Orientals say, "Greatness doth not approach him who is forever looking down; and all those who are looking high are growing poor."

It is remarkable that there is little or nothing to be remembered written on the subject of getting a living; how to make getting a living not merely honest and honorable, but altogether inviting and glorious; for if getting a living is not so, then living is not. One would think, from looking at literature, that this question had never disturbed a solitary individual's musings. Is it that men are too much disgusted with their experience to speak of it? The lesson of value which money teaches, which the Author of the Universe has taken so much pains to teach us, we are inclined to skip altogether. As for the means of living, it is wonderful how indifferent men of all classes are about it, even reformers, so called — whether they inherit, or earn, or steal it. I think that Society has done nothing for us in this respect, or at least has undone what she has done. Cold and hunger seem more friendly to my nature than those methods which men have adopted and advise to ward them off.

The title WISE is, for the most part, falsely applied. How can one be a wise man, if he does not know any better how to live than other men? — if he is only more cunning and intellectually subtle? Does Wisdom work in a treadmill? or does she teach how to succeed BY HER EXAMPLE? Is there any such thing as wisdom not applied to life? Is she merely the miller who grinds the finest logic? It is pertinent to ask if Plato got his living in a better way or more successfully than his contemporaries — or did he succumb to the difficulties of life like other men? Did he seem to prevail over some of them merely by indifference, or by assuming grand airs? or find it easier to live, because his aunt remembered him in her will? The ways in which most men get their living, that is, live, are mere makeshifts, and a shirking of the real business of life — chiefly because they do not know, but partly because they do not mean, any better.

The rush to California, for instance, and the attitude, not merely of merchants, but of philosophers and prophets, so called, in relation to it, reflect the greatest disgrace on mankind. That so many are ready to live by luck, and so get the means of commanding the labor of others less lucky, without contributing any value to society! And that is called enterprise! I know of no more startling development of the immorality of trade, and all the common modes of getting a living. The philosophy and poetry and religion of such a mankind are not worth the dust of a puffball. The hog that gets his living by rooting, stirring up the soil so, would be ashamed of such company. If I could command the wealth of all the worlds by lifting my finger, I would not pay such a price for it. Even Mahomet knew that God did not make this world in jest. It makes God to be a moneyed gentlemen who scatters a handful of pennies in order to see mankind scramble for them. The world's raffle! A subsistence in the domains of Nature a thing to be raffled for! What a comment, what a satire, on our institutions! The conclusion will be, that mankind will hang itself upon a tree. And have all the precepts in all the Bibles taught men only this? and is the last and most admirable invention of the human race only an improved muck-rake? Is this the ground on which Orientals and Occidentals meet? Did God direct us so to get our living, digging where we never planted — and He would, perchance, reward us with lumps of gold?

God gave the righteous man a certificate entitling him to food and raiment, but the unrighteous man found a facsimile of the same in God's coffers, and appropriated it, and obtained food and raiment like the former. It is one of the most extensive systems of counterfeiting that the world has seen. I did not know that mankind was suffering for want of gold. I have seen a little of it. I know that it is very malleable, but not so malleable as wit. A grain of gold will gild a great surface, but not so much as a grain of wisdom.

The gold digger in the ravines of the mountains is as much a gambler as his fellow in the saloons of San Francisco. What difference does it make whether you shake dirt or shake dice? If you win, society is the loser. The gold digger is the enemy of the honest laborer, whatever checks and compensations there may be. It is not enough to tell me that you worked hard to get your gold. So does the Devil work hard. The way of transgressors may be hard in many respects. The humblest observer who goes to the mines sees and says that gold digging is of the character of a lottery; the gold thus obtained is not the same thing with the wages of honest toil. But, practically, he forgets what he has seen, for he has seen only the fact, not the principle, and goes into trade there, that is, buys a ticket in what commonly proves another lottery, where the fact is not so obvious.

After reading Howitt's account of the Australian gold diggings one evening, I had in my mind's eye, all night, the numerous valleys, with their streams, all cut up with foul pits, from ten to one hundred feet deep, and half a dozen feet across, as close as they can be dug, and partly filled with water — the locality to which men furiously rush to probe for their fortunes — uncertain where they shall break ground — not knowing but the gold is under their camp itself — sometimes digging one hundred and sixty feet before they strike the vein, or then missing it by a foot — turned into demons, and regardless of each others' rights, in their thirst for riches — whole valleys, for thirty miles, suddenly honeycombed by the pits of the miners, so that even hundreds are drowned in them — standing in water, and covered with mud and clay, they work night and day, dying of exposure and disease. Having read this, and partly forgotten it, I was thinking, accidentally, of my own unsatisfactory life, doing as others do; and with that vision of the diggings still before me, I asked myself why I might not be washing some gold daily, though it were only the finest particles — why I might not sink a shaft down to the gold within me, and work that mine. THERE is a Ballarat, a Bendigo for you — what though it were a sulky-gully? At any rate, I might pursue some path, however solitary and narrow and crooked, in which I could walk with love and reverence. Wherever a man separates from the multitude, and goes his own way in this mood, there indeed is a fork in the road, though ordinary travelers may see only a gap in the paling. His solitary path across lots will turn out the higher way of the two.

Men rush to California and Australia as if the true gold were to be found in that direction; but that is to go to the very opposite extreme to where it lies. They go prospecting farther and farther away from the true lead, and are most unfortunate when they think themselves most successful. Is not our native soil auriferous? Does not a stream from the golden mountains flow through our native valley? and has not this for more than geologic ages been bringing down the shining particles and forming the nuggets for us? Yet, strange to tell, if a digger steal away, prospecting for this true gold, into the unexplored solitudes around us, there is no danger that any will dog his steps, and endeavor to supplant him. He may claim and undermine the whole valley even, both the cultivated and the uncultivated portions, his whole life long in peace, for no one will ever dispute his claim. They will not mind his cradles or his toms. He is not confined to a claim twelve feet square, as at Ballarat, but may mine anywhere, and wash the whole wide world in his tom.

Howitt says of the man who found the great nugget which weighed twenty-eight pounds, at the Bendigo diggings in Australia: "He soon began to drink; got a horse, and rode all about, generally at full gallop, and, when he met people, called out to inquire if they knew who he was, and then kindly informed them that he was 'the bloody wretch that had found the nugget.' At last he rode full speed against a tree, and nearly knocked his brains out." I think, however, there was no danger of that, for he had already knocked his brains out against the nugget. Howitt adds, "He is a hopelessly ruined man." But he is a type of the class. They are all fast men. Hear some of the names of the places where they dig: "Jackass Flat" — "Sheep's-Head Gully" — "Murderer's Bar," etc. Is there no satire in these names? Let them carry their ill-gotten wealth where they will, I am thinking it will still be "Jackass Flat," if not "Murderer's Bar," where they live.

The last resource of our energy has been the robbing of graveyards on the Isthmus of Darien, an enterprise which appears to be but in its infancy, for, according to late accounts, an act has passed its second reading in the legislature of New Granada, regulating this kind of mining; and a correspondent of the Tribune writes: "In the dry season, when the weather will permit of the country being properly prospected, no doubt other rich guacas [that is, graveyards] will be found." To emigrants he says: "Do not come before December; take the Isthmus route in preference to the Boca del Toro one; bring no useless baggage, and do not cumber yourself with a tent; but a good pair of blankets will be necessary; a pick, shovel, and axe of good material will be almost all that is required": advice which might have been taken from the "Burker's Guide." And he concludes with this line in italics and small capitals: "If you are doing well at home, STAY THERE," which may fairly be interpreted to mean, "If you are getting a good living by robbing graveyards at home, stay there."

But why go to California for a text? She is the child of New England, bred at her own school and church.

It is remarkable that among all the preachers there are so few moral teachers. The prophets are employed in excusing the ways of men. Most reverend seniors, the illuminati of the age, tell me, with a gracious, reminiscent smile, betwixt an aspiration and a shudder, not to be too tender about these things — to lump all that, that is, make a lump of gold of it. The highest advice I have heard on these subjects was groveling. The burden of it was — It is not worth your while to undertake to reform the world in this particular. Do not ask how your bread is buttered; it will make you sick, if you do — and the like. A man had better starve at once than lose his innocence in the process of getting his bread. If within the sophisticated man there is not an unsophisticated one, then he is but one of the devil's angels. As we grow old, we live more coarsely, we relax a little in our disciplines, and, to some extent, cease to obey our finest instincts. But we should be fastidious to the extreme of sanity, disregarding the gibes of those who are more unfortunate than ourselves.

In our science and philosophy, even, there is commonly no true and absolute account of things. The spirit of sect and bigotry has planted its hoof amid the stars. You have only to discuss the problem, whether the stars are inhabited or not, in order to discover it. Why must we daub the heavens as well as the earth? It was an unfortunate discovery that Dr. Kane was a Mason, and that Sir John Franklin was another. But it was a more cruel suggestion that possibly that was the reason why the former went in search of the latter. There is not a popular magazine in this country that would dare to print a child's thought on important subjects without comment. It must be submitted to the D.D.'s I would it were the chickadee-dees.

You come from attending the funeral of mankind to attend to a natural phenomenon. A little thought is sexton to all the world.

I hardly know an intellectual man, even, who is so broad and truly liberal that you can think aloud in his society. Most with whom you endeavor to talk soon come to a stand against some institution in which they appear to hold stock, that is, some particular, not universal, way of viewing things. They will continually thrust their own low roof, with its narrow skylight, between you and the sky, when it is the unobstructed heavens you would view. Get out of the way with your cobwebs; wash your windows, I say! In some lyceums they tell me that they have voted to exclude the subject of religion. But how do I know what their religion is, and when I am near to or far from it? I have walked into such an arena and done my best to make a clean breast of what religion I have experienced, and the audience never suspected what I was about. The lecture was as harmless as moonshine to them. Whereas, if I had read to them the biography of the greatest scamps in history, they might have thought that I had written the lives of the deacons of their church. Ordinarily, the inquiry is, Where did you come from? or, Where are you going? That was a more pertinent question which I overheard one of my auditors put to another once — "What does he lecture for?" It made me quake in my shoes.

To speak impartially, the best men that I know are not serene, a world in themselves. For the most part, they dwell in forms, and flatter and study effect only more finely than the rest. We select granite for the underpinning of our houses and barns; we build fences of stone; but we do not ourselves rest on an underpinning of granitic truth, the lowest primitive rock. Our sills are rotten.

What stuff is the man made of who is not coexistent in our thought with the purest and subtilest truth? I often accuse my finest acquaintances of an immense frivolity; for, while there are manners and compliments we do not meet, we do not teach one another the lessons of honesty and sincerity that the brutes do, or of steadiness and solidity that the rocks do. The fault is commonly mutual, however; for we do not habitually demand any more of each other.

That excitement about Kossuth, consider how characteristic, but super-ficial, it was! — only another kind of politics or dancing. Men were making speeches to him all over the country, but each expressed only the thought, or the want of thought, of the multitude. No man stood on truth. They were merely banded together, as usual one leaning on another, and all together on nothing; as the Hindus made the world rest on an elephant, the elephant on a tortoise, and the tortoise on a serpent, and had nothing to put under the serpent. For all fruit of that stir we have the Kossuth hat.

Just so hollow and ineffectual, for the most part, is our ordinary conversation. Surface meets surface. When our life ceases to be inward and private, conversation degenerates into mere gossip. We rarely meet a man who can tell us any news which he has not read in a newspaper, or been told by his neighbor; and, for the most part, the only difference between us and our fellow is that he has seen the newspaper, or been out to tea, and we have not. In proportion as our inward life fails, we go more constantly and desperately to the post office. You may depend on it, that the poor fellow who walks away with the greatest number of letters, proud of his extensive correspondence, has not heard from himself this long while.

I do not know but it is too much to read one newspaper a week. I have tried it recently, and for so long it seems to me that I have not dwelt in my native region. The sun, the clouds, the snow, the trees say not so much to me. You cannot serve two masters. It requires more than a day's devotion to know and to possess the wealth of a day.

We may well be ashamed to tell what things we have read or heard in our day. I do not know why my news should be so trivial — considering what one's dreams and expectations are, why the developments should be so paltry. The news we hear, for the most part, is not news to our genius. It is the stalest repetition. You are often tempted to ask why such stress is laid on a particular experience which you have had — that, after twenty-five years, you should meet Hobbins, Registrar of Deeds, again on the sidewalk. Have you not budged an inch, then? Such is the daily news. Its facts appear to float in the atmosphere, insignificant as the sporules of fungi, and impinge on some neglected thallus, or surface of our minds, which affords a basis for them, and hence a parasitic growth. We should wash ourselves clean of such news. Of what consequence, though our planet explode, if there is no character involved in the explosion? In health we have not the least curiosity about such events. We do not live for idle amusement. I would not run round a corner to see the world blow up.

All summer, and far into the autumn, perchance, you unconsciously went by the newspapers and the news, and now you find it was because the morning and the evening were full of news to you. Your walks were full of incidents. You attended, not to the affairs of Europe, but to your own affairs in Massachusetts fields. If you chance to live and move and have your being in that thin stratum in which the events that make the news transpire — thinner than the paper on which it is printed — then these things will fill the world for you; but if you soar above or dive below that plane, you cannot remember nor be reminded of them. Really to see the sun rise or go down every day, so to relate ourselves to a universal fact, would preserve us sane forever. Nations! What are nations? Tartars, and Huns, and Chinamen! Like insects, they swarm. The historian strives in vain to make them memorable. It is for want of a man that there are so many men. It is individuals that populate the world. Any man thinking may say with the Spirit of Lodin,

> "I look down from my height on nations,
> And they become ashes before me; —
> Calm is my dwelling in the clouds;
> Pleasant are the great fields of my rest."

Pray, let us live without being drawn by dogs, Esquimaux-fashion, tearing over hill and dale, and biting each other's ears.

Not without a slight shudder at the danger, I often perceive how near I had come to admitting into my mind the details of some trivial affair — the news of the street; and I am astonished to observe how willing men are to lumber their minds with such rubbish — to permit idle rumors and incidents of the most insignificant kind to intrude on ground which should be sacred to thought. Shall the mind be a public arena, where the affairs of the street and the gossip of the tea-table chiefly are discussed? Or shall it be a quarter of heaven itself — an hypethral temple, consecrated to the service of the gods? I find it so difficult to dispose of the few facts which to me are significant, that I hesitate to burden my attention with those which are insignificant, which only a divine mind could illustrate. Such is, for the most part, the news in newspapers and conversation. It is important to preserve the mind's chastity in this respect. Think of admitting the details of a single case of the criminal court into our thoughts, to stalk profanely through their very sanctum sanctorum for an hour, ay, for many hours! to make a very barroom of the mind's inmost apartment, as if for so long the dust of the street had occupied us — the very street itself, with all its travel, its bustle, and filth, had passed through our thoughts' shrine! Would it not be an intellectual and moral suicide? When I have been compelled to sit spectator and auditor in a courtroom for some hours, and have seen my neighbors, who were not compelled, stealing in from time to time, and tiptoeing about with washed hands and faces, it has appeared to my mind's eye, that, when they took off their hats, their ears suddenly expanded into vast hoppers for sound, between which even their narrow heads were crowded. Like the vanes of windmills, they caught the broad but shallow stream of sound, which, after a few titillating gyrations in their coggy brains, passed out the other side. I wondered if, when they got home, they were as careful to wash their ears as before their hands and faces. It has seemed to me, at such a time, that the auditors and the witnesses, the jury and the counsel, the judge and the criminal at the bar — if I may presume him guilty before he is convicted — were all equally criminal, and a thunderbolt might be expected to descend and consume them all together.

By all kinds of traps and signboards, threatening the extreme penalty of the divine law, exclude such trespassers from the only ground which can be sacred to you. It is so hard to forget what it is worse than useless to remember! If I am to be a thoroughfare, I prefer that it be of the mountain brooks, the Parnassian streams, and not the town sewers. There is inspiration, that gossip which comes to the ear of the attentive mind from the courts of heaven. There is the profane and stale revelation of the barroom and the police court. The same ear is fitted to receive both communications. Only the character of the hearer determines to which it shall be open, and to which closed. I believe that the mind can be permanently profaned by the habit of attending to trivial things, so that all our thoughts shall be tinged with triviality. Our very intellect shall be macadamized, as it were, its foundation broken into fragments for the wheels of travel to roll over; and if you would know what will make the most durable pavement, surpassing rolled stones, spruce blocks, and asphaltum, you have only to look into some of our minds which have been subjected to this treatment so long.

If we have thus desecrated ourselves — as who has not? — the remedy will be by wariness and devotion to reconsecrate ourselves, and make once more a fane of the mind. We should treat our minds, that is, ourselves, as innocent and ingenuous children, whose guardians we are, and be careful what objects and what subjects we thrust on their attention. Read not the Times. Read the Eternities. Conventionalities are at length as bad as impurities. Even the facts of science may dust the mind by their dryness, unless they are in a sense effaced each morning, or rather rendered fertile by the dews of fresh and living truth. Knowledge does not come to us by details, but in flashes of light from heaven. Yes, every thought that passes through the mind helps to wear and tear it, and to deepen the ruts, which, as in the streets of Pompeii, evince how much it has been used. How many things there are concerning which we might well deliberate whether we had better know them — had better let their peddling-carts be driven, even at the slowest trot or walk, over that bridge of glorious span by which we trust to pass at last from the farthest brink of time to the nearest shore of eternity! Have we no culture, no refinement — but skill only to live coarsely and serve the Devil? — to acquire a little worldly wealth, or fame, or liberty, and make a false show with it, as if we were all husk and shell, with no tender and living kernel to us? Shall our institutions be like those chestnut burs which contain abortive nuts, perfect only to prick the fingers?

America is said to be the arena on which the battle of freedom is to be fought; but surely it cannot be freedom in a merely political sense that is meant. Even if we grant that the American has freed himself from a political tyrant, he is still the slave of an economical and moral tyrant. Now that the republic — the res-publica — has been settled, it is time to look after the res-privata — the private state — to see, as the Roman senate charged its consuls, "ne quid res-PRIVATA detrimenti caperet," that the private state receive no detriment.

Do we call this the land of the free? What is it to be free from King George and continue the slaves of King Prejudice? What is it to be born free and not to live free? What is the value of any political freedom, but as a means to moral freedom? Is it a freedom to be slaves, or a freedom to be free, of which we boast? We are a nation of politicians, concerned about the outmost defenses only of freedom. It is our children's children who may perchance be really free. We tax ourselves unjustly. There is a part of us which is not represented. It is taxation without representation. We quarter troops, we quarter fools and cattle of all sorts upon ourselves. We quarter our gross bodies on our poor souls, till the former eat up all the latter's substance.

With respect to a true culture and manhood, we are essentially provincial still, not metropolitan — mere Jonathans. We are provincial, because we do not find at home our standards; because we do not worship truth, but the reflection of truth; because we are warped and narrowed by an exclusive devotion to trade and commerce and manufactures and agriculture and the like, which are but means, and not the end.

So is the English Parliament provincial. Mere country bumpkins, they betray themselves, when any more important question arises for them to settle, the Irish question, for instance — the English question why did I not say? Their natures are subdued to what they work in. Their "good breeding" respects only secondary objects. The finest manners in the world are awkwardness and fatuity when contrasted with a finer intelligence. They appear but as the fashions of past days — mere courtliness, knee-buckles and smallclothes, out of date. It is the vice, but not the excellence of manners, that they are continually being deserted by the character; they are cast-off clothes or shells, claiming the respect which belonged to the living creature. You are presented with the shells instead of the meat, and it is no excuse generally, that, in the case of some fishes, the shells are of more worth than the meat. The man who thrusts his manners upon me does as if he were to insist on introducing me to his cabinet of curiosities, when I wished to see himself. It was not in this sense that the poet Decker called Christ "the first true gentleman that ever breathed." I repeat that in this sense the most splendid court in Christendom is provincial, having authority to consult about Transalpine interests only, and not the affairs of Rome. A praetor or proconsul would suffice to settle the questions which absorb the attention of the English Parliament and the American Congress.

Government and legislation! these I thought were respectable professions. We have heard of heaven-born Numas, Lycurguses, and Solons, in the history of the world, whose names at least may stand for ideal legislators; but think of legislating to regulate the breeding of slaves, or the exportation of tobacco! What have divine legislators to do with the exportation or the importation of tobacco? what humane ones with the breeding of slaves? Suppose you were to submit the question to any son of God — and has He no children in the Nineteenth Century? is it a family which is extinct? — in what condition would you get it again? What shall a State like Virginia say for itself at the last day, in which these have been the principal, the staple productions? What ground is there for patriotism in such a State? I derive my facts from statistical tables which the States themselves have published.

A commerce that whitens every sea in quest of nuts and raisins, and makes slaves of its sailors for this purpose! I saw, the other day, a vessel which had been wrecked, and many lives lost, and her cargo of rags, juniper berries, and bitter almonds were strewn along the shore. It seemed hardly worth the while to tempt the dangers of the sea between Leghorn and New York for the sake of a cargo of juniper berries and bitter almonds. America sending to the Old World for her bitters! Is not the sea-brine, is not shipwreck, bitter enough to make the cup of life go down here? Yet such, to a great extent, is our boasted commerce; and there are those who style themselves statesmen and philosophers who are so blind as to think that progress and civilization depend on precisely this kind of interchange and activity — the activity of flies about a molasses-hogs-head. Very well, observes one, if men were oysters. And very well, answer I, if men were mosquitoes.

Lieutenant Herndon, whom our government sent to explore the Amazon, and, it is said, to extend the area of slavery, observed that there was wanting there "an industrious and active population, who know what the comforts of life are, and who have artificial wants to draw out the great resources of the country." But what are the "artificial wants" to be encouraged? Not the love of luxuries, like the tobacco and slaves of, I believe, his native Virginia, nor the ice and granite and other material wealth of our native New England; nor are "the great resources of a country" that fertility or barrenness of soil which produces these. The chief want, in every State that I have been into, was a high and earnest purpose in its inhabitants. This alone draws out "the great resources" of Nature, and at last taxes her beyond her resources; for man naturally dies out of her. When we want culture more than potatoes, and illumination more than sugar-plums, then the great resources of a world are taxed and drawn out, and the result, or staple production, is, not slaves, nor operatives, but men — those rare fruits called heroes, saints, poets, philosophers, and redeemers.

In short, as a snowdrift is formed where there is a lull in the wind, so, one would say, where there is a lull of truth, an institution springs up. But the truth blows right on over it, nevertheless, and at length blows it down.

What is called politics is comparatively something so superficial and inhuman, that practically I have never fairly recognized that it concerns me at all. The newspapers, I perceive, devote some of their columns specially to politics or government without charge; and this, one would say, is all that saves it; but as I love literature and to some extent the truth also, I never read those columns at any rate. I do not wish to blunt my sense of right so much. I have not got to answer for having read a single President's Message. A strange age of the world this, when empires, kingdoms, and republics come a-begging to a private man's door, and utter their complaints at his elbow! I cannot take up a newspaper but I find that some wretched government or other, hard pushed and on its last legs, is interceding with me, the reader, to vote for it — more importunate than an Italian beggar; and if I have a mind to look at its certificate, made, perchance, by some benevolent merchant's clerk, or the skipper that brought it over, for it cannot speak a word of English itself, I shall probably read of the eruption of some Vesuvius, or the overflowing of some Po, true or forged, which brought it into this condition. I do not hesitate, in such a case, to suggest work, or the almshouse; or why not keep its castle in silence, as I do commonly? The poor President, what with preserving his popularity and doing his duty, is completely bewildered. The newspapers are the ruling power. Any other government is reduced to a few marines at Fort Independence. If a man neglects to read the Daily Times, government will go down on its knees to him, for this is the only treason in these days.

Those things which now most engage the attention of men, as politics and the daily routine, are, it is true, vital functions of human society, but should be unconsciously performed, like the corresponding functions of the physical body. They are infra-human, a kind of vegetation. I sometimes awake to a half-consciousness of them going on about me, as a man may become conscious of some of the processes of digestion in a morbid state, and so have the dyspepsia, as it is called. It is as if a thinker submitted himself to be rasped by the great gizzard of creation. Politics is, as it were, the gizzard of society, full of grit and gravel, and the two political parties are its two opposite halves — sometimes split into quarters, it may be, which grind on each other. Not only individuals, but states, have thus confirmed dyspepsia, which expresses itself, you can imagine by what sort of eloquence. Thus our life is not altogether a forgetting, but also, alas! to a great extent, a remembering, of that which we should never have been conscious of, certainly not in our waking hours. Why should we not meet, not always as dyspeptics, to tell our bad dreams, but sometimes as eupeptics, to congratulate each other on the ever-glorious morning? I do not make an exorbitant demand, surely.

Dilemmas Without Technical Solutions

A number of problems that face human populations are problems without technical or scientific solutions.* These problems may be compared to the child's game of tic-tac-toe, for here too is a problem with no solution; no one can win at tic-tac-toe if his opponent is knowledgeable and alert. This is why tic-tac-toe *is* a child's game.

What Hardin has called the "tragedy of the commons" illustrates the types of problems confronting mankind that lack logical solutions. The "commons" is the meadow on which all citizens of a village graze their cattle. It can easily support a certain number of animals, but if the number were to be increased considerably, the milk and beef production of each animal and of the entire herd would drop. Overcrowding might possibly lead to the demise of the herd.

Despite the threat of overgrazing and the possible loss of all cattle, individual cattle men have no reason not to add additional animals to their herds. For the individual herdsman, adding one animal yields a profit nearly as large as the contribution of any one animal he already owns. The loss that arises from overgrazing is shared by all and so it affects the individual herdsman but slightly.

The above analysis is true for each herdsman, and so each adds one animal after another until the commons is ruined for all. Ironically, the addition of each animal to each herd appears at the moment to be a profitable step to the individual herdsman but, all the time, the entire system moves inexorably towards destruction. The same dismal pattern applies to the growth of human populations and to man's pollution of his environment. It also applies to the armaments race between powerful nations that refuse to negotiate with other equally powerful ones except from a position of superiority. If each insists on being "superior," the result is merely an unending escalation of destructive armaments.

The tragedy of the commons runs its predestined course because the problem has been stated in terms of individual herdsmen who act with no consideration for others, as if they live in isolation. The problem can be diagrammed as follows:

*See G. Hardin, "The Tragedy of the Commons," *Science, 162* (1968), 1243-1248.

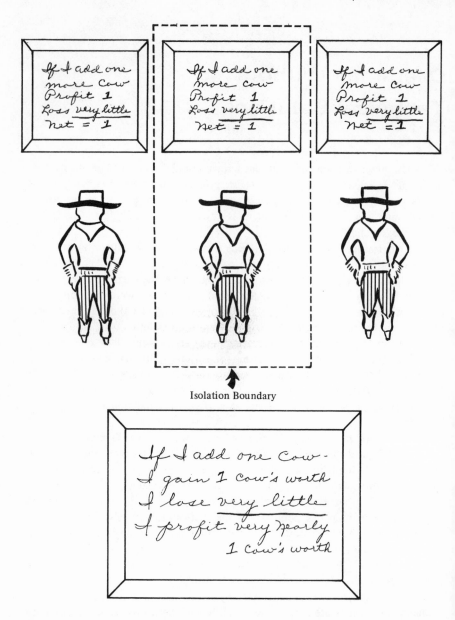

Isolation Boundary

The dashed line isolates the single herdsman as well as the reasoning he uses in evaluating the consequences of adding an additional animal to his herd. The individual herdsman's problem is solved by merely adding gains and losses to find the net profit.

The fatal error of the obvious arithmetic solution diagrammed above lies in the assumed independence of individual herdsmen; far from being independent, these men and their families are all members of a village, of a community of interdependent citizens. If the village food supply collapses, each herdsman and his family will suffer. The diagram representing the true situation, therefore, must be made with connecting lines between the herdsmen. These lines represent the bonds of society.

The number of cattle that yields the greatest profit for all equals 97

Isolation Boundary

The dashed line in the second diagram is unable to cut the community bonds and, consequently, is forced to include the entire village; in this case, the mathematical analysis of cattle grazing is based on the combined herds of the village.

A problem such as that of the commons, of population growth, or of environmental pollution has no technical solution when viewed from the standpoint of the isolated individual. According to Professor Hardin, problems of this sort require a fundamental extension of morality. As our diagrams make clear, morality in this sense has its origins in the interrelations between persons that prevent the isolation of a single individual from the community of which he is part, and the examination of the problem solely from his private (and incomplete) point of view. The connecting links between individuals reflect the interdependence of these individuals upon one another in a human community.

In large measure the problems that face mankind in respect to numbers of individuals, food production, pollution, and preservation of the environment are not soluble from the individual's limited view — not from the point of view of the individual person, the individual industry, the individual nation, *or even the individual species.* The dashed line cannot be drawn across the interconnecting links if correct answers are to ensue from logical analyses of man's perplexing problems. This will be a recurrent theme of many subsequent essays.

On the Growing Role
of Technology

Ours is a technological society.* Let those who doubt this claim or who fail to grasp its meaning consider the space program that put two men on the moon in 1969. Let them consider the bank of computers and consoles and the swarm of workers at the Houston Space Center. Let them imagine the contracts, the subcontracts, and the sub-subcontracts that produced the flow of hardware that was finally assembled as space capsules, rockets, and rocket launchers. Was there a town of 50,000 inhabitants anywhere in the United States that did not have a small manufacturing firm that was party to one of these ramifying contracts? Or is there in the United States today any city of similar size that has been overlooked by contracts let by the Defense Department for military equipment and supplies?

On a somewhat more modest scale, he who only poorly understands the role of technology in our society might consider a single plant owned by but one chemical company, a plant engaged in the production of elemental phosphorus for fertilizers (less than 40 per cent of the total putput) and household detergents (over 60 per cent!). Each year this plant produces nearly 300 million pounds of phosphorus, which is shipped directly in molten form to processing plants. The amount of electrical energy needed by this one plant for its yearly output is 1.7 billion kilowatt hours – an amount equal to 2 per cent of the entire electrical output of the Tennessee Valley Authority.

A technological society can operate only if its parts mesh smoothly. Large numbers of errors in production (internal errors) cannot be tolerated. In order to assure that they do not occur, both persons and machines are, of necessity, worked much below the limits of their capacity. For workers, white- or blue-collar, this means that they must be under-used; the simplistic, routine steps performed at each station along a factory's assembly line are simplistic and routine by design.

In our society attitudes toward the increasing role of technology and, in turn, science differ sharply. To many, technology has brought man uncounted blessings – plentiful food, freedom from disease, a longer and more varied

*For detailed, contrasting views on the role of technology in society, see John McDermott, "Technology, the Opiate of the Intellectuals," *New York Review, 31* (1969) and Harvard University Program on Technology and Society, *Fourth Annual Report, 1967-1968*, Cambridge, Mass. (1968), reviewed by McDermott.

life – and promises still more, including eventual peace on earth, goodwill toward men. Some of these technological optimists see that these presumed blessings have been delivered together with a packet of undesirable side effects such as pollution of the environment, overpopulation, and social upheavals. Nevertheless, they believe that the undesirable consequences of technology (its *negative externalities* as opposed to *positive opportunities,* which are the desired outputs or goals of the technological system) will be solved by further technology. Technology, in the eyes of its ardent supporters, is a self-correcting system that operates for the good of society. Because of the similarity of this view to the doctrine espoused by industrialists of the last century *(laissez faire),* McDermott names the doctrine of technologists *laissez innover* – in fractured French, "Let's innovate."

McDermott is a very convincing technological pessimist. Technology is not self-correcting in respect to negative externalities because these unwanted side effects are nonexistent within the frame of reference of technology's managers; on the contrary, they exist only in the eyes of those who are outside the technological system. To those within the system, a positive opportunity – to place a man on the moon in this decade, to deliver 300,000 tons of high explosives to the battle front monthly, to produce 300 million pounds of phosphorus or 1 billion pounds of pesticides annually – governs all feeder processes of the manufacturer's flow chart. To keep the entire operation running smoothly, doubts and misgivings within the managerial hierarchy must be weeded out ("I'm sick of hearing about this damned report – write the goddamned thing and shut up about it" – testimony before a congressional hearing as reported in the *New York Times,* August 14, 1969). Since the goal (the positive opportunity) is rational, opposition to it must be irrational. Furthermore, because technological systems tend to increase in scale and complexity, because they spread to incorporate new areas and to incorporate these into ever-growing systems, and because they adopt increasingly ambitious goals, they become increasingly resistant to the recognition of or response to negative externalities. The positive opportunity becomes the sole end; all else is excluded from the calculations and, as a result, self-correction as an attribute of technology becomes a myth.

The arguments presented by McDermott explain the vehemence with which pesticide manufacturers have attacked Rachael Carson and others who have opposed the indiscriminate spraying of millions of tons of long-lived poisons into the atmosphere, onto the land, and into the waters of the nation. They explain the vigor with which officials of the oil industry attack conservationists who oppose offshore drilling, the pollution of vast areas of the sea, and the destruction of forested areas by the construction of pipelines and roadways. They explain the obvious petulance of transportation officials at outspoken opposition to the unlimited construction of jetports and superhighways. In each instance, those within the system look at the limited goal set for their technological system – set in many instances by computers that have been programmed to regard last year's output as "good" and its transcendence as "better" — as *the* rational one; surely, then, those in opposition to the overall goal must be "kooks," "longhairs," or "commies."

As for their effects on society, according to McDermott, because decisions as to *rational* and *irrational* are made by the managerial class working entirely within the technological system, these decisions at best are neutral in respect to what is rational or irrational in respect to society as a whole.

Is technology basically evil? Is it a mark of sanity to argue for a return to the arts-and-crafts society of medieval Europe? Could we turn back to a simpler life even if we wished to? McDermott's views of technology lead him to a pessimistic outlook on the future of society. He sees the conflicting demands on persons caught up in the technological system — more training but relatively less use made of it, increasing stratification of society but a need for greater educational opportunities for all — leading to sharp conflicts within society.

The force of the description of technology and how it operates is, in my opinion, much in McDermott's favor. I appreciate his jibe at the claim that at no time in history has such a high proportion of people *felt* like individuals; a claim so preposterous reflects an oblivion to a decade of student unrest on the nation's campuses.

Does it follow, therefore, that open conflict within society is a preordained fate of the technological system? Here is my opinion: That more technology, as we now know it, will provide a solution for negative externalities of already established technological procedures strikes me as a futile hope. As with a person struggling in quicksand, technology left to solve its own problems, reasoning from its own self-contained frame of reference, will surely multiply mankind's problems and make their solutions even more difficult to attain. The secret must lie (and I use "must" in the sense of both "obligation" and "to the exclusion of other possibilities") in enlarging the positive opportunities of technology to include the preservation in perpetuity of the finite resources of a finite world. For too long technologists have pretended that factories, mines, and dams — although individually large (a source of local pride) — are infinitesimally small in contrast to the limitless earth on which they are placed. Similarly, technologists have pretended for too long that their slag heaps, their fumes, their heat, and even their final products are negligible in quantity, and that the world is a trash basket of infinite size. Too long technologists have estimated that one day, one year, or 50 years is "forever." Radioactive wastes, some of which have a half-life of 24,000 years, were buried in containers designed to last for 50 years; for the early containers, that time is now half gone. Nor are technologists the only ones at fault. For too long people — individual persons, the church, political organizations, and other social groups — have looked upon the surface of the earth as an infinite plain upon which persons can be housed without limit and from which nourishment for unlimited numbers of persons can be drawn.

Nowhere in McDermott's otherwise excellent and inclusive review do I sense that he sees the smallness of the earth as a serious matter. The conflict over the rationality or irrationality of technological procedures comes from conservationists and their colleagues who appreciate the extent to which technology is exhausting and poisoning the earth's surface in opposition to industrialists and their spokesmen who either do not appreciate these adverse effects or consider them to be unimportant in comparison with more limited positive opportunities. Referring to the prologue of this book, I feel that McDermott has concentrated on the coming conflicts between passengers or between passengers and crew *given existing conditions*. He shows no evidence of having thought seriously of ending the "Game," of restoring the plundered ship, or of discouraging the arrival of new passengers.

The solution to problems raised by a technological society will come from education; I see no other possibility. I would guess that persons in government may prove to be those most easily convinced of the larger goals to which technological systems must address themselves. In part this guess is based on hindsight because even today a considerable number of lawmakers, especially at the federal level, are taking firm stands in opposition to the uncontrolled destruction of the nation's resources and to the unconcerned exploitation of a technologically naive population. I would guess, too, that knowledgeable young engineers, technologists, and executives in sufficient number may break the resistance of the older managerial personnel to admitting the existence of negative externalities. Although technology may be able to weed out the sporadic espousal of enlightened causes as the rantings of irrational persons, it will be unable to withstand the persistent and frequent espousals by large numbers of its lower managerial echelons because it is from these young persons that the industrial leaders of tomorrow must be chosen.

If I were not to believe in the power of education or if I were to lack faith in the collective sense of today's students, I would be well advised to leave the university for a less exacting, more remunerative position. If I were to find myself booking passage aboard the *Titanic,* as Paul Ehrlich has expressed it, I would insist on going first-class.

Number of Persons:
The Inexorable Exponential*

Two centuries before the birth of Christ, Archimedes, a Greek mathematician, claimed that he could count the grains of sand that would be required to fill the entire space contained within the universe. At that time the largest named number was the "myriad" (10,000). To obtain still larger numbers, Archimedes devised a notation that allowed him to manipulate myriads-of-myriads and still higher orders of numbers. In effect, Archimedes devised an exponential scheme corresponding to the present-day notations 10, 10^2 (100), 10^3 (1000), 10^4 (10,000), 10^5 (100,000), and so on. Numbers multiplied by themselves, specifically large numbers multiplied by themselves, yield exceedingly large products. For example, if every one of the 100,000 spectators at an Army-Navy football game were to construct his own stadium capable of holding 100,000 spectators, the entire population of the world (4 billion persons) would not fill the 10 billion stadium seats that would now be available ($10^5 \times 10^5 = 10^{10}$).

The rapidity with which events ultimately change under exponential growth has been known since early times. An ancient Indian fable tells of the wise man whose requested payment for a favor rendered his king was merely the grains of wheat necessary to cover a chess board; that is, 1 grain on the first square, 2 grains on the second, 4 grains on the third, 8 grains on the fourth, and so on for all 64 squares. The amount of wheat requested under this seemingly modest proposal was equal to the entire wheat crop of India for 2,000 years.

Exponential increases are deceptive; events move slowly, nearly imperceptibly, for a long period and then proceed with frightening speed. The Midwestern farmer does not ask that wheat rust not attack his crop because this would be asking too much; he asks only that his wheat ripen and be harvested before the rust passes through an appreciable number of generations. Seven rust generations might be required before the fungus can spread from a single infected plant to 10 acres of a 1,000-acre ranch, but only two additional generations might be required to wipe out the remaining 990 acres. The last two generations, consequently, would wipe out 99 per cent of the crop, with the last generation alone causing 90 per cent of the total damage.

This phrase is taken from the title of Philip H. Abelson's "The Inexorable Exponential," *Science,* 162 (1968), 221.

The perceptive wheat farmer will adopt almost any procedure that prolongs the generation time of this fungal disease.*

The growth of populations tends to be exponential; parents usually leave more offspring than two and so, until limited by lack of food, water, or living space, a population grows at an ever-increasing (exponential) rate. The ceiling imposed from the outside may be reached with startling speed. A population of *Drosophila* flies will reach the limit of nearly any laboratory population cage within two generations because one female can lay as many as 500 to 700 eggs. The 300 first-generation daughters of a single initial female can produce 150,000 to 210,000 offspring – many more than the 10,000 to 20,000 flies that an ordinary laboratory cage can support. A comparable limit was previously described in discussing the spread of wheat rust: 1,000 acres of wheat would be completely infected after nine rust generations, provided that each infected plant released enough rust spores to infect 10 other plants; the spread of a real infection may be much more rapid than this.

Rational persons wish to avoid the imposition of population control on man by means of environmental limitations because lack of food and lack of housing exert their effects largely through the death and starvation of children. Unfortunately, these same rational persons are not necessarily familiar with the characteristic nature of exponential growth. They do not realize that in many instances trouble is not detected until it is too late to act. A novice wheat farmer who was unfamiliar with wheat rust would not realize that less time remained between the spread of an infection from 10 acres to 1,000 acres than from 1 plant to 10 acres. He might never understand that if he could not respond quickly enough by harvesting the remaining mature grain, the loss of 10 acres of infected wheat was virtually identical to the loss of an entire crop of 1,000 acres.

The tragedy of amateurish incompetence can be illustrated by a fictional (but not unrealistic) presidential commission that is sent to examine the population problems of various states within the United States. Before the investigation gets under way, the inhabitants of the nation have been redistributed in the following manner (see table on page 55): one person is placed within the smallest state, Rhode Island; two persons within the next larger state, Delaware; four within the next larger, Connecticut; and so forth. This pattern, it will be recalled, is that of the wheat grains on the squares of the chess board.

The president instructs his commission to visit the states sequentially by increasing size and to report to him on a state-by-state basis the conditions as members of the commission interpret them; he is especially interested in learning of any state that has reached a population level of 200 to 400 persons per square

*This topic is also treated in the essay entitled "The Arithmetic of Epidemics" in Volume III.

A Fictional Redistribution of Persons Within the United States
Illustrating the Abrupt Change from Tolerable to Intolerable
Population Densities Between the 24th and 26th States

State	Area (1,000 sq mi)	Population	Persons/sq mi
1. Rhode Island	1	1	0.001
2. Delaware	2	2	0.001
3. Connecticut	5	4	0.001
4. Hawaii	6	8	0.001
5. New Jersey	8	16	0.002
6. Massachusetts	8	32	0.004
7. New Hampshire	9	64	0.007
8. Vermont	10	128	0.013
9. Maryland	11	256	0.023
10. West Virginia	24	512	0.021
11. South Carolina	31	1,024	0.033
12. Maine	33	2,048	0.062
13. Indiana	36	4,096	0.114
14. Kentucky	40	8,192	0.205
15. Virginia	41	16,384	0.400
16. Ohio	41	32,768	0.799
17. Tennessee	42	65,536	1.560
18. Pennsylvania	45	131,072	2.913
19. Mississippi	48	262,144	5.461
20. Louisiana	49	524,288	10.700
21. New York	50	1,048,576	20.972
22. Alabama	52	2,097,152	40.330
23. North Carolina	53	4,194 304	79.138
24. Arkansas	53	8,388,608	158.276
25. Wisconsin	56	16,777,216	299.593
26. Iowa	56	33,554,432	599.186
27. Illinois	56	67,008,864	1198.372
28. Michigan	58	[134,017,728]	[2310.650]

mile, for he understands that these are the population densities of China and
India — countries that appear to be overcrowded.

If we judge from official reactions to the population problem of the real
world, the tragedy of the commission's work is that the well-meaning but
"realistic" commissioners would go from Rhode Island, to Delaware, to
Connecticut, to Hawaii, and to other states in order of their size, sending back to
the White House in each case an entirely optimistic report: "No evidence of
overpopulation whatsoever; reports of impending tragedy grossly misleading."
These optimistic reports would continue through the first 23 states visited; the
population density of the 23rd state would be 79 persons per square mile
(4,194,304 persons living within the 53,000 square miles of North Carolina).

Uncertainty, perhaps consternation, might strike the commission in the 24th state, in which the population density would be 158 persons per square mile, still lower than the president's "worry level." The report of the 25th state, however, would be devastating: 300 persons per square mile. The lesson to be learned is one that has been repeated several times in this essay: quantities that increase exponentially mislead the uninformed; ultimate limits are reached with fantastic speed. The attitude of our hypothetical commission changed from an unworried complacency based on studies of 23 states to concern in the 24th to shock in the 25th. If they were to continue their inspection trip, they would find nearly half of the population of the United States crowded into Michigan (the 28th state in order of size) at a density greater than 2,300 persons per square mile.

Should the reader be disconcerted by the imaginary travels of an investigative commission from one state to another under the highly artificial conditions described here, he might consider the reassuring reports that many real investigative commissions and learned committees have made public while the world moves from one year to another and from one generation to another. We have just now entered the generation in which consternation has become widespread; we are living, so to speak, in the 24th state. And the succession of generations presses on as inexorably as the exponential itself.

On the Conservation of Matter

These paragraphs are a reminder of a law with which we are all familiar, at least in its trivial applications, the conservation of matter. If two substances are sealed into a glass tube and then allowed to react with one another (by heating the tube perhaps), the total weight of the tube and its contents remains unchanged even though the reaction may involve a change of the substances in the tube from one physical state to another, from a liquid to a solid or from a solid to a gas.

Although the conservation of matter may be appreciated intuitively at the laboratory level, it is almost universally ignored at the level of everyday living. Where do autumn leaves go when they are thrown onto a bonfire? Why, they burn up. Where does the tank full of gasoline go on a weekend drive? It is burned up in the engine. Where do new cars go when they leave Detroit? To satisfied customers, of course. Ah, but what about the cars that were previously owned by those customers? They are resold. Eventually, however, they accumulate in large piles at the edges of cities and towns.

In our daily lives we forget that matter does not disappear and we forget that we live in a world of finite size. In the past human beings could overlook these facts; it is no longer possible to overlook them because there are now 4 billion persons. Four billion is an enormously large number; by the use of fly swatters alone, 1 billion Chinese have virtually exterminated the house fly in mainland China!

Four billion persons in the course of their lives transform tremendous quantities of materials from one form to another — usually to an unwanted pollutant. No one who has approached New York City by jet from the West Coast and has seen the reddish-brown dome of metropolitan smog extending 200 miles to Harrisburg, Pennsylvania, can think of the earth's atmosphere as an infinite receptacle for man's waste. Nor can we continue to ignore the waste produced by transcontinental jets themselves.

Throughout his existence man has buried his wastes in the ground, spilled them into rivers and streams, or allowed them to rise into the air. As long as his wastes were largely organic wastes and as long as he existed in small numbers, the disposal of waste matter in this way did not harm his environment. Organic waste eventually succumbs to the action of soil animals and microbes and, having been broken down to simple compounds, reenters the web of life once more, largely through the action of green plants.

The problems of pollution arise from the quantities of waste that are now produced because of the large numbers of persons and from the types of wastes that are now discarded. The organic waste produced by 200 million Americans is tremendous; urine, feces, and garbage leave the homes by pipes and trucks while poultry farms, slaughterhouses, and vegetable processers seek inexpensive ways of disposing of industrial garbage. If a nearby lake seems to be a reasonable dumping place, we quickly learn to our dismay that only the merest trace of nitrogenous wastes stagnating for a single summer will choke the lake with unwanted and unpleasant algal and bacterial growth.

Nonorganic matter confronts us with unique problems. Many of the substances now thrown onto the ground and into the lakes and rivers are not degradable by bacteria. Detergents find their way into and accumulate in ground and surface water; to this our foamy rivers attest. Persistent pesticides and herbicides are sprayed onto the soil year after year in ever increasing amounts. Those substances accumulate not only because by design they are "persistent" but also because they destroy much of the microflora and microfauna that are responsible for ecological transformations of organic materials in the soil.

Periods of time that at a given moment appear long to an engineer working under pressure to solve a particular problem seem absurdly short to the rest of us in retrospect. At the first International Conference on the Peaceful Uses of Atomic Energy held in Geneva, Switzerland, in 1955, engineers described the disposal of long-lived radioactive wastes by burial in steel containers that would last 50 years. Fifty years? That may have seemed like an eternity when the urgent need to dispose of these wastes first arose, but half of the life expectancy of the early containers has now passed; the problem of the disposal of radioactive wastes was merely postponed then, and it has not been adequately solved even now.

Man solves his big problems by breaking them into smaller parts that he can handle rather easily. This is the procedure, however, that led to the "tragedy of the commons" described in the previous essay. It is the unwarranted oversimplification at the level of individual industries of problems encompassing all of society that makes pollution of the environment seem not only tolerable but profitable. Furthermore, an abysmal ignorance of the conservation of matter prompts many sincere persons to believe not only "out of sight, out of mind" but also "out of sight, out of existence." Automobiles account for most of the air pollution in the United States. In groping for a solution to this problem, no industry spokesman has yet spoken of fewer cars; few spokesmen have suggested smaller cars with smaller engines. Most suggestions from the automobile industry have favored a false solution — converting pollutants into still smaller particles so that they cannot be seen. Together, faulty procedural analyses and ignorance of the basic law of the conservation of matter account for much of the dangerous optimism in today's world.

The Water Budget

Life and water are inseparable. Living tissue is largely water. The chemical reactions necessary for life are aqueous reactions. Body fluids of living things so nearly resemble sea water in their composition that the very similarity suggests that primeval life arose in the sea. The dependence of life upon water is so great that vast arid regions such as much of the Sahara Desert may be devoid of living things.

The absolute limit imposed on life by the amount of available water enables us to calculate from the total rainfall the largest number of persons that can inhabit a given area, provided that we know (at least approximately) their standard of living.

The average rainfall throughout the United States is about 30 inches per year. The total area of the country is about 2 billion acres (3 million square miles) and so the total rainfall is about 4,300 billion gallons per day. Of this amount, 1,300 billion gallons run off as rivers and streams or contribute to the ground water; the remainder, 3,000 billion gallons, evaporates directly from the soil or is used by and transpired from plants.

The amount of runoff water that is used daily in the United States has been estimated at one-quarter of the 1,300 billion gallons. Metered water that is sold to the consumer amounts to 240 billion gallons daily, but the same water can be sold again and again of course; cities along a single river use the same water repeatedly as it is withdrawn for one city's water supply, returned through the local sewage system, withdrawn farther downstream for the next city's drinking water, returned again, and so on. Many cities do not sell metered water (New York City is one), and so the 240 billion gallons of metered water is an underestimate of the amount actually used.

Two types of evidence suggest that there is not a great deal more usable water in the 1,300 billion-gallon daily runoff. First, the rivers and streams of the country are about as choked with waste – industrial and human – as can be tolerated. All of the great rivers have suffered this fate. There is a limit to the concentration of waste products in water that is to be used for drinking and cooking.

The other evidence of America's insufficient supply of usable runoff water is our overuse of ground water. Ground water is a portion of the runoff water. Too much of it is being used by cities and for irrigation in the West, the Northeast, the Midwest, and the Middle Atlantic states. We find, as a result, a continual drop in the water table throughout most of the United States. In effect, "capital" is being used in an effort to maintain existing communities and industries and our current standard of living.

The most severe restriction on the largest possible size of the population of the United States arises not from runoff water but from transpired water. This is the water that produces our food. The wheat for 2½ pounds of bread requires 300 gallons of water. Much more water is required to produce a pound of beef. Each day a steer will eat 30 pounds of alfalfa and drink 12 gallons of water. Because 800 pounds of water (100 gallons) are required to produce each pound of alfalfa, the daily water requirement of a steer is 3,000 gallons. A two-year-old steer provides about 700 pounds of beef, and so each pound is equivalent to the use of 3,000 gallons of water. One person's daily diet, which at one extreme may consist of a mere loaf of bread or at the other include a pound of beef, requires for its production anything from 300 to 3,000 gallons of water.

Man's need for water exceeds that required for food alone because much of his clothing and the furnishings of his home is also obtained from living things; a pound of cotton, for example, requires a great deal of water. To include all of these additional needs in our calculations, we may estimate that the 3,000 billion gallons of transpired water in the United States are used almost entirely for the production of man's crops and for meeting his other needs. For the average American this represents a daily average water requirement of 15,000 gallons. It is quite obvious that the water budget of the United States is very nearly exhausted by the 200 million persons living as they now live. The 3,000 gallons required for a pound of beef simply is not that far removed from the 15,000 gallons arrived at by subdividing the entire amount of transpired water. Man does not live by beef alone; there is not a great deal of surplus to absorb either greater numbers of persons or greater demands per person.

On the drawing boards there exist plans for nuclear reactors that will convert sea water to fresh water. One such reactor is designed to produce 620 million gallons of fresh water as well as 1.5 million kilowatts of electricity per day. Can we assume that reactors of this sort represent a reasonable solution to the problems facing America's water budget?

The transformation of sea water into fresh water by desalinization does not offer a long-range solution to the water problem in a country such as the United States. A number of vexing problems exist that seem to have been largely neglected while work has proceeded on the design of the reactor itself. First, a great deal of heat will be produced; in a later essay thermal pollution will be discussed in detail. Here I shall simply say that the most obvious method to remove the heat — water-cooling of the reactor — will alter the temperature and the character of the sea for large distances around each nuclear plant.

The second problem lies in the 100,000 tons of salt that will accumulate each day as the by-product of desalinization. Salt, as well as water, is part of sea water, and when pure water is removed, the salt remains. Presumably the salt would be returned to the sea in the form of hot, concentrated brine. The nearby sea, therefore, will not only be heated (thermal pollution) but will also have an exceptionally high salt concentration (industrial pollution). The ocean near the site of each nuclear reactor promises to become a Dead Sea in every sense of the term.

The third problem associated with any nuclear reactor is that of disposing of the long-lived radioactive isotopes that are produced during the reactor's operation. Only a few choices are possible: bury them, throw them in the sea, or let them rise into the air. For large reactors, air and water can no longer serve as cesspools; the isotopes must be buried. In time, however, they will threaten the nation's ground water because the metal containers will disintegrate.

These three major problems have not yet been given sufficient thought — neither to solve them nor even to appreciate fully their magnitude. Perhaps in time suitable solutions can be found by those responsible for reactor design and construction. There remains, however, an additional problem that at the moment appears to be fatal to the entire idea of supplying the water needs of the United States by nuclear-operated desalinization plants. The 620 million gallons that such a plant might produce daily would be sufficient with our present standard of living for the total needs of only 50,000 persons. These needs are not restricted to drinking, washing, and flushing toilets but also include the water needed for the production of meat and vegetables and for the wood and fibers used by people in the United States. But recently the population of this country has been growing by more than 10,000 persons per day, and this rate will increase. Thus, the design problems that have taken years to solve and the thermal, brine, and isotope-disposal problems that have yet to be resolved adequately (even for a single reactor) can be traced to a need to provide water for persons equal in number to just a five-day increase in the country's population. To construct every five days one such desalinization plant and to face its concomitant problems is utterly absurd. The solution to the problems of the water budget lies in the control of numbers of persons, not in searching for sources of water.

The Air We Breathe:
An Essay in Two Parts

1. On the gain and loss
of atmospheric oxygen

This is an essay about oxygen, the life-supporting portion of the earth's atmosphere. It emphasizes the role of green plants (plants that possess chlorophyll) in making oxygen available to us. The oxygen cycle in the living world illustrates on a massive scale the conservation of matter. The cycle also illustrates the extremely narrow margin within which man and other living things operate. To make these points clear requires little more than elementary arithmetic.

The atmosphere surrounding the earth is immense. Its total weight is 5×10^{15} tons, of which one-fifth or 10^{15} (1 quadrillion or 1 followed by 15 zeros) tons are oxygen. The bulk of what is left is nitrogen, an inert gas that plays no direct role in respiration. The total weight of the atmosphere has been calculated (and can be verified by the reader if he wishes) by multiplying the earth's surface (196,938,800 square miles) by air pressure at sea level (14.7 pounds per square inch).

The enormous quantity of oxygen in the earth's atmosphere today exists because of green plants. These plants — and only they — are capable of carrying out the following chemical reaction:

$$\text{Carbon dioxide} + \text{Water} + \text{Energy from sunlight} \rightarrow \text{Starch} + \text{Oxygen}$$

The reaction could have been written to emphasize either the numbers of molecules taking part in the reaction (six molecules of carbon dioxide, six molecules of water, one molecule of starch, and six molecules of oxygen) or the relative weights of the reacting compounds (264 grams of carbon dioxide, 108 grams of water, 180 grams of starch, and 192 grams of oxygen; each side of the equation adds up to 372 grams in this latter case).

The starch that is made by the plant takes part in a series of subsequent metabolic reactions (both within the plant and within animals that eat plants) that release energy in the form of heat. The green plant, in effect, converts light energy into heat. The release of heat is accomplished in many small steps, but the overall reaction is merely the reverse of the one given earlier:

$$\text{Starch} + \text{Oxygen} \rightarrow \text{Carbon dioxide} + \text{Water} + \text{Heat}$$

When an animal eats a plant, when a fungus grows on another plant, or when an animal eats another animal, all of the energy that keeps the herbivore, the saprophytic plant, or the carnivore alive comes from chemical reactions that fit within the restriction written directly above. The source of energy that keeps life on earth going is sunlight captured by green plants and stored in a chemical (starch) from which it can be released as heat when needed.

Oxygen in pure gaseous form is a relative newcomer to the earth's environment. Before green plants arose about a billion years ago in the evolution of life, oxygen existed mainly in stable compounds such as carbon dioxide (CO_2), water (H_2O), and metallic oxides (rusts such as iron oxide). Other compounds that were common in the primeval atmosphere were methane (CH_4, swamp gas) and hydrogen cyanide (HCN, a deadly poison to life today). The 10^{15} tons of oxygen that are now found in the atmosphere have accumulated over the past billion years; it comes as a byproduct of the formation of starch from carbon dioxide and water. The average rate of accumulation can be calculated by dividing 10^{15} tons by 10^9 years; it equals 10^6 tons per year. An average of 1 million tons of oxygen have accumulated each year in the earth's atmosphere through the action of the world's (green) plant life.

The amount of oxygen produced annually by plants greatly exceeds the average amount that has accumulated in the atmosphere each year; in a comparable fashion, the annual income of a family greatly exceeds the yearly savings that accumulate in the bank. The annual production of oxygen by all plant life on earth has been estimated at 400 billion tons: 90 per cent of this comes from marine plankton; 10 per cent, from terrestrial plants. The accumulation of 1 million tons of oxygen in the atmosphere represents little more than two-millionths or so of the total annual production. If the annual accumulation is looked upon as an annual saving, it represents a very small fraction of the total income indeed — a yearly saving of 2 cents for a family whose annual income is $10,000.

What happens to the 400 billion tons of oxygen that is produced annually? It is transformed back into carbon dioxide and water when living organisms (including green plants themselves) burn starch for energy. The small amount of oxygen that dribbles into the atmosphere (small, but terribly important) can be accounted for in the following way:

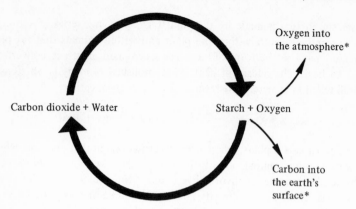

Oxygen into
the atmosphere*

Carbon dioxide + Water   Starch + Oxygen

Carbon into
the earth's
surface*

   *For every 32 pounds of oxygen that escape into the atmosphere, 12 pounds of carbon escape unoxidized to form coal, oil, natural gas, or other organic carbon in the earth's surface.

The oxygen that remains in the atmosphere as a surplus product of photosynthesis remains there only because unburned carbon has accumulated on the earth's crust or under the sea. The oxygen in the air corresponds directly to the unburned carbon, to coal, oil, oil shale, natural gas, graphite, and diamonds. Both the carbon and the free oxygen exist as such because they happened to escape from the worldwide carbon cycle. In general, the escape was made possible because organic material accumulated where the oxygen level was too low to support the bacteria that would otherwise decompose such material into carbon dioxide and water. Under proper geologic conditions and in the continued absence of oxygen, organic material is converted into one of the fossil fuels, the main source of energy for modern civilization.
   It should be clear that man's burning of fossil fuels for industrial purposes represents a reintroduction of carbon into the carbon cycle. As carbon is removed from the store of unoxidized fuels and is burned, a corresponding amount of oxygen is removed from the atmosphere. Indeed, if the entire 4×10^{14} tons of carbon of organic origin present on earth were to be burned, the 10^{15} (1 million billion or 1 quadrillion) tons of oxygen in the atmosphere would be entirely depleted too. Are we in danger of doing just this?
   The total energy obtained annually from fossil fuels in the United States equals 48×10^{15} BTU (British thermal units, the amount of heat that is required to raise the temperature of one pound of water one degree Farenheit); 36×10^{15} BTU are obtained by burning petroleum products and the remaining 12×10^{15} BTU from burning coal. Coal, when it is burned, releases 12,000 BTU per pound; a pound of oil releases about 20,000 BTU. By dividing, we can calculate that 12^{12} pounds of coal and 1.8×10^{12} pounds of oil are burned annually; the sum equals 2.8×10^{12} pounds or about 1.4×10^9 tons of carbon. The amount of oxygen needed to burn carbon is 2 2/3 times the weight of the carbon; thus, 3.8 billion tons of oxygen are required annually to run the industries and homes of the United States.

The 3.8×10^9 tons of oxygen required annually for industrial purposes seem to be but a small fraction of the 10^{15} tons of oxygen that exist in the atmosphere; it represents only 1 per cent of the total amount of oxygen produced each year. It greatly exceeds, however, the average amount of oxygen that has been added annually to the earth's atmosphere for millennia. An average of 1 million tons have been added each year to the world's atmosphere, and the United States now removes nearly 4 billion tons annually.

Within the oxygen budget, therefore, we are at this moment operating on capital. That in itself is not necessarily bad; retirement funds and annuities are designed to provide incomes that utilize both capital and interest. The tacit agreement in such schemes, however, is that the benefactor pass away at a time that can be predicted with reasonable accuracy. No one has made such an agreement for mankind; no one has suggested that we should cease to exist. Therefore, the matter of living on capital needs careful scrutiny.

To start, we should realize the consequences that would occur if the population of the entire world were to attain the standard of living enjoyed within the United States. The population of the world is 4 billion and of the United States 200 million; Americans are but one-twentieth of the present world population. If the entire world were to use energy at the rate we use it in the United States, the atmosphere would lose 7.6×10^{10} tons of oxygen each year in burning 2.8×10^{10} tons of carbon. This amounts to nearly 20 per cent of the world's output of oxygen! It is not an amount that can be treated lightly. The average annual accumulation of 1 million tons of oxygen is negligible compared to the 76 billion tons of oxygen that an industrial world would remove from the atmosphere. The question is reduced to how long the accumulated oxygen in the air will last.

What fraction of the total oxygen content of the air can be removed with impunity? Certainly a 10-percent decrease in oxygen tension at sea level would be a serious matter for heavy smokers and for those who suffer from heart or respiratory ailments. Cities at high altitudes, such as Denver, Mexico City, and Bogotá, would be so seriously affected that they would be abandoned. Worse than that, however, is the possible death of many green plants when the oxygen pressure is lowered by 10 per cent; many plant species would die. Such a reduction in oxygen might easily start the world's oxygen supply on a downward spiral, in the face of which only anaerobes (some microbes that use certain organic compounds other than free oxygen for their energy source) would survive. Should the amount of oxygen in the atmosphere be lowered by as much as 1 per cent, I would imagine that mankind would be on the brink of an unprecedented disaster. And 1 per cent of the total oxygen of the atmosphere equals only 10^{13} tons.

How long would it be before the industrial giants of the world burned 10^{13} tons of oxygen? It is impossible to say because we do not know precisely when the underdeveloped nations will be heavily industrialized. We can say, however, that if every person on earth were to use energy at the rate an average citizen of the United States uses it today, it would require 10^{13} tons divided by

8×10^{10} tons per year or 125 years to burn up 1 per cent of the earth's total oxygen supply. This is a short time. Indeed, the actual time before 1 per cent is consumed may be considerably shorter than 125 years because since 1930 the demand for energy in the United States has increased 10-fold per 30 years. Within 30 years, therefore, the United States alone will be using one-half of the amount we have calculated as the total demand for the entire world. If the industrialization of underdeveloped nations succeeds, the worldwide demand for oxygen might be five or ten times that which we used in our calculations. Two conclusions seem inescapable. First, oxygen is disappearing from the atmosphere because of our enormous appetite for energy. Second, if the rest of the world attempts to emulate the United States in the use of energy, serious trouble may arrive in less than a century — possibly in several decades.

Addenda:

In 1968 Professor Lamont Cole published an essay entitled "Can the World Be Saved?"* I was aware of his essay (it is not easy to forget the alternative title: "Is there intelligent life on earth?") but used little or none of it in preparing the preceding account of the gain and loss of atmospheric oxygen. In the meantime, two articles have been published in which Dr. Cole is taken to task for his concern over the possible loss of oxygen through some calamitous man-made event.† Dr. Broecker refers to the prospect of oxygen loss as a "bogeyman"; Dr. Ryther comments on "doomsday prophecy" and "atmosphere of eco-catastrophe."

Upon reading the two recent articles, my instinctive reaction was to revise and rework my own essay to include the points of view expressed by these more knowledgeable authors. On second thought, however, I have decided to let the original essay remain as it was written. Left unchanged, it serves as an excuse for examining diverse points of view; that is, it illustrates the plea I made in the preface to this book: "read the essays..., enjoy them, react to them, and through prolonged discussion criticize and *rethink* them." Evidence that an author can err may be more valuable to the student under some circumstances than the information that would be conveyed by a surreptitiously edited essay.

When two sincere men, having concerned themselves with a common problem, arrive at contradictory conclusions, their disagreement must be based

*Lamont Cole, "Can the World Be Saved?" *Bioscience, 18* (1968), 679-684.
†John H. Ryther, "Is the World's Oxygen Supply Threatened?" *Nature, 227* (1970), 374-375.
Wallace S. Broecker, "Man's Oxygen Reserves," *Science, 168* (1970), 1537-1538.
See also Lamont Cole's "Oxygen Reserves," *Environment, 12* (1970), 40-42.
Several letters concerning man's oxygen supply also appear in this issue of *Environment*.

on one of the following reasons. (1) The factual information available to them differed and, consequently, one of them, perhaps both, has been misinformed. (2) The two in fact, have been considering separate facets of what was erroneously thought to be a single problem. (3) Even though their data were identical and the immediate problem truly one, complicating and complicated circumstances have caused the two men to react differently and thus to reach different conclusions. The case of the disappearing oxygen offers examples of all of these.

The one fact over which there is little or no disagreement is the amount of oxygen in the earth's atmosphere. Both the air pressure at sea level and the surface area of the earth are known; from these two items we can calculate the total amount of air. The composition of the air is homogeneous for a height of at least 20, perhaps 60, miles; oxygen contributes about 21 per cent by volume or (because the molecular weight of oxygen is 16 and that of nitrogen only 14) about 23½ per cent by weight. In my essay I used 20 per cent as the proportion of oxygen in the atmosphere even though I was discussing weight. Had I been more precise, I would have calculated that the atmosphere contains 1.2×10^{15} tons of oxygen, an amount identical to Ryther's estimate. Broecker gives the amount of oxygen in terms of moles per square meter of the earth's surface (1 mole of oxygen equals 32 grams) and says that there are 60,000 moles per square meter. Cole, in his reply, says there are 74,000 moles per square meter, an amount in closer agreement with a total weight of 1.2×10^{15} tons.

The small differences in the quoted amounts of oxygen in the atmosphere are trivial in the sense that any careful elementary-school student can perform the calculations needed to arrive at the correct answer. More serious is the variation in the estimate of the relative contributions of marine and land plants to the total production of oxygen. I used values obtained from a text by Eugene I. Rabinowitch: 90 per cent, marine; and 10 per cent, land.* Cole, in his original article, says that 70 per cent or more of the oxygen produced by photosynthesis is produced in the oceans. Ryther, in contrast, claims the proportions are nearer 30 per cent marine to 70 per cent land, whereas Broecker suggests nearly the reverse: 60 per cent, marine; and 40 per cent, land. Thus, among the more divergent estimates are those of Dr. Cole's two critics; experimental evidence on this issue is apparently still imprecise. This, as I have said, is one reason why sincere men can differ.

What about the total production of oxygen? I claimed that some 400 billion tons are produced by photosynthesis annually; the amount that can be calculated from Rabinowitch's text is 400 to 530 billion tons. Ryther's views on total oxygen production can be calculated as follows. He says that the net production of oxygen by marine plants equals 150 million tons annually;

*Eugene I. Rabinowitch, *Photosynthesis*, Vol. 1 (New York: Interscience Publishers, 1945).

therefore, because the sea produces (in his view) only one-third of all oxygen, the net world production would equal 450 million tons. He also estimates that only $\frac{1}{1,000}$ of the oxygen produced annually remains as net production; hence, the total world production would be about 450 billion tons annually.

Broecker's estimate of total yearly production of oxygen must be reconstructed from his estimate of an average production of 8 moles per square meter or, using his figures, $\frac{8}{60,000}$ of the total amount in the atmosphere. This amounts to only some 150 billion tons, one-third the two estimates given above. Cole estimates the total annual production of oxygen as 1.43×10^{17} grams or, approximately, 150 billion tons; thus, his figure is nearly identical to Broecker's.

In my essay I calculated an *average* rate of oxygen accumulation in the atmosphere by dividing 10^{15} tons of atmospheric oxygen by 10^9 years (the approximate time that green plants have existed); the average obtained in this way is 1 million tons per year. Cole believes that the oxygen level has remained relatively constant for the past half-billion years; I would have to admit, therefore, that the rate of accumulation was once higher than it is now and that today fewer than 1 million tons of oxygen are accumulating per year. My estimate, probably the least reliable of three, is much smaller than those of the others. We should bear in mind, however, that *accumulation in the atmosphere* need not be identical to *net production by photosynthesis;* should there be a loss of oxygen by means other than the oxidation of organic material, the two values could be substantially different. Ryther says that the net production of oxygen by the sea alone is 150 million tons annually; this suggests that the net world production (three times as great) would be 450 million tons. Broecker claims that the net production is $\frac{1}{15,000,000}$ the total amount of atmospheric oxygen or about 60 million tons. Cole, because he believes that the organic processes that produce and consume oxygen are in equilibrium, would claim that there is no net production of oxygen at the present time.

At this point I shall digress to consider a point emphasized by Cole. It seems that the amount of oxygen in the atmosphere has not changed by 1 part in 1,000 during the past 60 years. If there are 10^{15} tons of oxygen in the atmosphere, then the total has not increased or decreased by as much as 1,000 billion (that is, 1 trillion) tons during 60 years or by 15 billion tons per year. Should the net production be as either Ryther, Broecker, or I have calculated and should this net production accumulate each year in its entirety, the change in the total amount of oxygen would be thousands of times too small to be detected by man's measuring devices. Similarly, the current rate of oxygen loss through the combustion of fossil fuels is still too small to be detected readily.

Rabinowitch, in his book on photosynthesis, said that photosynthesis by green plants alone prevents the rapid disappearance of all life from the face of the earth. Cole sounded the same theme in his essay by saying that oxygen would disappear from the atmosphere through natural geological processes if all the marine plants were to be killed. Ryther calculates that the sea is responsible for the annual net production of about 150 million tons of oxygen. Suppose marine plants were destroyed so that these 150 million tons were no longer contributed annually to the atmosphere. Suppose, too, that the amount of oxygen in the atmosphere is constant only because poorly understood losses counterbalance this annual contribution. What length of time would be required before the annual loss of 150 million tons would be felt? If there are 12×10^{14} tons of oxygen and if 15×10^7 tons were to be lost each year, then after 1 million years 15×10^{13} tons or approximately 10 per cent of the earth's atmospheric oxygen would have disappeared. One million years is a long time. Nor are matters speeded up immensely by arguing that nearly all of the oxygen is really produced by marine plants because, at best, this would shorten the time required for the loss of 10 per cent of the atmospheric oxygen to some 300,000 years. Still a long time! The amount of oxygen that has accumulated in the atmosphere is by now sufficiently great (that is, it has accumulated over a sufficiently long time) that, should the annual input by photosynthesis in the ocean cease, the world of life on land could continue to run for some time. But the process would be one way and irreversible!

The problem posed by Ryther as a result of Cole's statement on the destruction of marine plants (photosynthetic diatoms) is not the problem I posed in my essay on the gain and loss of atmospheric oxygen. I admitted that the net annual production of oxygen is infinitesimal compared to the amount being consumed by the combustion of fossil fuels. My concern lay with the oxidation of carbon that has been sequestered in and under the earth's surface during the past billion years, a sequestering that was necessary for the accumulation of atmospheric oxygen. My calculations suggest that nearly 4×10^9 tons of oxygen are burned annually for industrial and other purposes; this amount is small in comparison with the total of 10^{15} tons but it is thousands of times larger than that which I calculated as the average annual gain from photosynthesis. My concern, therefore, is for the depletion of oxygen through the burning of fossil fuels, and it is to this problem that Broecker addresses himself.

Broecker states his position precisely: "If we were to burn all known fossil fuel reserves we would use less than 3 per cent of the available oxygen. Clearly a general depletion of the atmospheric oxygen supply via the consumption of fossil fuels is not possible in the foreseeable future."

Reactions to Broecker's casual attitude toward the effect of burning fossil fuels can be expressed in the form of three questions. How reliable is his estimate of fossil fuel reserves? Is it possible that man's activities will create demands for oxygen that *must be added to* those created by the burning of fossil fuels? Should one dismiss so lightly the loss of 3 per cent (or of any other substantial amount) of our one and only oxygen supply?

In my essay I made the mathematically correct statement that *if* we were to burn all carbon of organic origin, we would simultaneously burn all oxygen in the earth's atmosphere. How much carbon of organic origin, however, exists as fossil fuel? Broecker apparently settles for 3 per cent. Cole, on the other hand, points out that this value is constantly being revised upward and, taking the new Alaskan oil fields into account, it should be nearer 8 per cent. I am not at all sure whether either of these figures includes reliable estimates of the oil and gas to be found off the coasts of Africa or under the China Sea and North Sea. Do these figures allow for new finds in Asia, Africa, or Antartica? Do they allow for the eventual commercial use of oil shales and tar sands, two enormous reserves of sequestered carbon not yet tapped?

Cole raises an additional point concerning the withdrawal of oxygen from the atmosphere through the oxidation of reduced chemicals in the earth's surface layers. One loss of this sort would occur, for example, through the weathering of tropical soils if these were cultivated extensively in the future. A second loss would occur in the oxidation of materials turned up (literally plowed up) during strip mining and strip-mining-like procedures used in man's increasingly frantic search for rare minerals.

Even more serious, in my estimation, is Broecker's complete neglect of events that may be set in motion by seemingly trivial changes in the environment. We can claim that a 3-percent or a 10-percent loss of oxygen can be tolerated because man (a healthy man) does not die (at sea level) if his oxygen supply is reduced by such an amount. More precisely, he does not drop dead at once as he would if a small amount of carbon monoxide were to be added to the air he breathes.

The suggestion that the loss of a small proportion of oxygen from the atmosphere is harmless is fraught with danger. First of all, the loss decreases the range of altitudes over which human populations can exist; it drives mankind into the valleys of the world. To speak lightly of the loss of 3 per cent of the atmospheric oxygen is to speak lightly of native tribes living at high altitudes in the Andes and the Himalayas, human beings whose fate can be sealed by anonymous persons living unseen beyond the farthest horizon.

Moreover, if we admit that man is but one thread in an interconnected skein of life on earth, we should not limit our immediate concern to mankind alone. Species do not often disappear through the sudden death of their individual members; the relative reproductive rates of different species determine which will continue to exist and which will disappear from the earth. Trivial changes in the salinity or oxygen content of water in an aquarium often drastically alter the composition of its community of microrganisms. In later essays in this section I shall contrast the stability of natural systems before certain onslaughts with their amazing vulnerability to other attacks. In one particular case I can say that a mere handful of inconspicuous beetles protects 3 million acres of California rangeland. I am positive in this case because these beetles are the few remaining descendants of an enormous population (itself started by the introduction of small numbers of imported parents) that cleared the rangeland of the Klamath weed, which in the 1930's had rendered the range useless for grazing. The Klamath weed still exists in California and so do small numbers of the beetles that live on them but the complementary roles of these two species in respect to the agricultural economy of California and its neighboring states could never be guessed from their present-day pitiful numbers.

Background complexities of this sort are not included in Broecker's reasoning when he confidently argues that a mere 3 per cent of our oxygen will have disappeared together with our fossil fuel reserves; presumably his confidence would persist if the loss of oxygen amounted to 8 per cent or 10 per cent instead. On the other hand, I shudder at the possible changes in the composition of life about us that might be initiated by a 1-per cent change in atmospheric oxygen. To be optimistic in the face of man's almost complete ignorance of ecological interactions is, in my estimation, to indulge in brinksmanship – to play Russian roulette, as I explain in a later essay, according to the strategies designed for pocket billiards.*

2. Carbon dioxide and life on earth

The bulk of the earth's atmosphere consists of the two gases, nitrogen and oxygen – about 80 per cent of the former and 20 per cent of the latter. These proportions are only approximate, of course, for there are many minor gaseous components of air, including, for example, argon and water vapor. Also among the minor components is carbon dioxide, the gas used by green plants to manufacture starch. The remarkable, almost miraculous ability of plants to utilize carbon dioxide becomes clear when we realize that it represents only 0.05 per cent of air (5 parts in 10,000); to manufacture one pound of starch, a plant must process more than one ton of air.

Because the atmosphere contains only a very small amount of carbon dioxide (about 2,500 billion tons in all), the production of large quantities of this gas by industry through the burning of coal, oil, and gas threatens to bring about substantial changes in its *relative* concentration in the earth's atmosphere. Besides serving as raw material in the synthesis of starch, what does carbon dioxide do as a component of our atmosphere? In this brief essay I shall consider two of its effects. First, it plays an important role in trapping at the earth's

*See the essay entitled "Brinksmanship: The Limits of Game Theory" in this volume.

surface heat that arrives as infrared radiation from the sun; this is the "greenhouse" effect of carbon dioxide. Second, it plays an additional role in determining the acidity of seawater; the carbon dioxide in the air is in equilibrium with a much larger amount that is dissolved (as carbonic acid) in water and that is important in determining the amounts and kinds of dissolved salts present in seawater.

The warmth of a greenhouse on a sunny day depends on the difference in wavelengths of infrared (heat) waves as they come to us from the sun and as they are reflected from the floor and benches within the greenhouse. In the sense that infrared rays have longer wavelengths than do those of light rays that make up the visible portion of electromagnetic radiation, all radiant heat has long wavelengths. Nevertheless, the radiation that arrives from the sun has relatively short wavelengths and, as a consequence, can penetrate the atmosphere and other optically clear substances such as glass. The radiant heat that is reflected from terrestrial surfaces, however, has much longer wavelengths. Glass reflects rather than transmits these long waves and, hence, heat is trapped within a greenhouse or any other glass-enclosed room exposed to sunlight. Similarly, carbon dioxide in the atmosphere reflects the long heat waves back toward the surface of the earth, thereby trapping much of the heat that arrives from the sun. Carbon dioxide in the atmosphere operates in the same manner as the panes of glass in a greenhouse; it operates as a heat trap.

An increase in the amount of carbon dioxide in the earth's atmosphere promises to increase the temperature at the earth's surface. No one has predicted the rise in temperature that might follow an increase in atmospheric carbon dioxide but that the temperature will rise is not in dispute. Opinions differ in respect to the change in climate that might follow a world wide temperature rise. An abnormal melting of polar ice has been suggested as one possibility, an effect that might raise the level of the world's oceans and inundate many continental coastal plains. On the other hand, a higher humidity in temperate zones, an increased wintertime snowfall, and the eventual formation of newly glaciated areas have been suggested as an alternative series of events that might follow the initial warming of the earth. The extreme suggestion, one whose final details are only of academic interest to man, states that the greenhouse effect could lead to temperatures high enough to boil water. The increased water vapor would then add to the total greenhouse effect, thus increasing the temperature even more until the cloud cover formed by water vapor condensing at high altitudes cuts down the incident radiant heat at the earth's surface. Under this extreme view, increasing the carbon dioxide in the atmosphere might tip matters so that an entirely new and different meteorological equilibrium would be established but only after the earth had become very unlike the world we know today. Even without the world being tipped into a drastically new equilibrium, the ecology of a warmed earth would eventually come to differ greatly from that within which man now lives.

The effect of an increase of atmospheric carbon dioxide on the acidity of both fresh water and seawater has been the subject of considerable theoretical calculations. The effect of altered acidity (measured as pH, the negative logarithm of the hydrogen ion concentration) on marine life under laboratory conditions is well known to experimental biologists. Marine fish (herring for example) can tolerate only minute alterations in the acidity of seawater. The sperm of many types of marine organisms fail to fertilize eggs if the pH of the water into which they are released differs from that of seawater. Although lower organisms such as plankton, diatoms, mollusks, and others tolerate greater changes in pH than do most higher organisms, these lower forms also require strikingly rigid conditions.

In the case of many organisms, a 50-percent increase in the total amount of atmospheric and oceanic carbon dioxide could create difficult if not insurmountable hardships. I say this not only on the basis of the expected alteration of the pH of seawater but also on the assumption that this alteration would lead in turn to a decrease in the amount of dissolved calcium carbonate and to alterations in the amounts and proportions of other salts that are essential for life in the sea.

The total carbon dioxide contained in the air and waters of the earth amounts to a staggering 70,000 billion (7×10^{13}) tons. At the rate American industry consumes coal, oil, and gas, we introduce about 5 billion tons of carbon dioxide into the air annually. The first reaction, therefore, is to claim that a 50-percent increase in the present amount of carbon dioxide would require some 7,000 years; at least, that is the answer obtained by dividing 35,000 billion tons by 5 billion tons per year. This first estimate should be adjusted, however, by assuming that industrialization − the American Way of Life − will soon spread to all peoples on earth. In this case, the total annual production of carbon dioxide would increase 20-fold and the time required to produce one-half the amount now found in the air and seas (should our supply of fossil fuels prove sufficient) is reduced to 350 years. Lest this seems like an enormous interval of time, recall that the United States celebrates its 200th birthday during the 1970's.

The estimations given in the paragraph immediately preceding can be revised drastically (and justifiably) by recalling that most marine life occurs in the shallow seawater above the various continental shelves; surface water in the center of large oceans and the water in the ocean depths are surprisingly devoid of life. It is also necessary to recall that there is only a very slow turnover of water from the depths of the ocean, to its surface, and back again. In this case, the amount of carbon dioxide in the air and waters of the earth might be *effectively* increased by one-half after the world's industrial production of this gas equaled half of the total found in the atmosphere plus that which is dissolved in a surface layer of the ocean 500 feet (roughly one-tenth of a mile) deep. Whether or not this adjustment is a realistic one will depend upon the length of time required for industrial carbon dioxide to reach the necessary amounts. If only decades are required, the alteration of carbon dioxide levels dissolved in surface waters might easily prove to be a serious problem.

The solution to the problem as we have now rephrased it hinges on the proportion of all oceanic waters that is contained in a layer one-tenth of a mile deep, a depth not unlike that of the relatively shallow waters above the continental shelves. The total surface area of the world's oceans amounts to 135 million square miles; a layer one-tenth of a mile deep would contain 13.5 million cubic miles. The total volume of water on earth is about 300 million cubic miles; therefore, less than 5 per cent of the total volume of all water occurs in the relatively thin, one-tenth-mile surface layer. Now, if 350 years would be required for a heavily industrialized world to contaminate all of the ocean's waters with dangerous levels of carbon dioxide (half again of the total amount already present), only 20 years would be required to contaminate the surface layers for a depth of 500 feet to an equally dangerous extent. That the estimated time, 20 years, should be so short suggests that the dissolved carbon dioxide would in fact be largely limited to the upper layers because enormously long time periods are needed to thoroughly mix the waters of an ocean.

Still another aspect of the carbon dioxide problem needs discussion before the matter is concluded. Green plants use carbon dioxide to make starch that in turn supplies the metabolic energy needed for plant growth. As carbon dioxide increases in the atmosphere and in the oceans' waters, would it not follow that the amount of green plant life would also increase, thus removing carbon dioxide from the air and converting it into living material? Yes – and no. Green land plants compete with man for living space and, as human populations have swelled in size, forest lands have all but disappeared. Even jungles in remote areas are not as lush and widespread as elementary-school textbooks would lead us to believe; man is encroaching upon plant life in all parts of the globe. One cannot argue, therefore, that terrestial plant life will cope with the additional atmospheric carbon dioxide.

Plant life in the oceans still accounts for most of the oxygen production that occurs on the earth. Because of the rate at which noxious industrial pollutants are now being poured into the ocean, there are many who predict that within a few decades the oceans will suffer the same fate that the Great Lakes suffered during the 1960's – they will stagnate and die. Remember that the bulk of oceanic plant life occurs in the relatively shallow and easily polluted water above the offshore continental shelves. If bacteria prove to be the most hardy living things in a deteriorating environment, they will decompose once-living organic matter, thus completing the carbon cycle and again releasing the carbon dioxide.

So the withdrawal of carbon dioxide from the air and water by green plants does not provide an effective escape from the dangers posed by the ever-increasing production of this gas by the growing and spreading industries of the world. Rampant industrialization is incompatible with the continuation of life on earth as we know it, incompatible with the continued existence of lower forms of life, and incompatible with man's existence as well.

The Potato Famine:
Population Control Gone Wrong

The years from 1846-1851 were years of famine in Ireland. The potato crop failed. One million persons perished; more than 1.5 million others emigrated, largely to the United States. This tragedy has a lesson for us: populations do not move smoothly toward equilibrium conditions and then come comfortably to rest. The number of inhabitants that a nation can support with its available food is not arrived at automatically. Populations change in size by fits and starts; curves illustrating these changes are jagged, not smooth. The potato famine of Ireland is an irregularity on a graph. What caused it? Can it happen again — for example, in the United States?

Between 1779 and 1841, the population of Ireland increased from a rather stable 3 million to 8 or perhaps 9 million persons. This increase must be identified as the ultimate cause of the famine, as the reason why 2.5 million persons either died in Ireland or left home to avoid starvation.

Many factors are responsible for explosive increases of human populations. The means for increase, in the sense of physical ability to produce children in large numbers, is always present; women have been known to bear children at one- or two-year intervals for 15 or 20 years or more. In stable populations this ability either is not used or its use is counterbalanced by an extremely high infant and child mortality.

The potato was responsible in large part for the growth of Ireland's population — that is, the potato and, because of it, the young age at which the Irish of the last century married. Ireland's potatoes provided a much greater yield per acre than did the grain that the English and Scots used as their staple food. A family could easily grow a year's supply of potatoes on a small plot of ground. Young couples — sixteen or eighteen years of age — needed very little else to start married life than that small plot for raising potatoes. What were once sizable family holdings were split and resplit to accommodate ever-growing numbers of children and grandchildren.

The Irish couples of the early 19th century were prolific; they married young, had their first child almost immediately, and had later ones thereafter at regular one- or two-year intervals. Now, the exponential growth of a population is extremely sensitive to generation time. To illustrate this fact, suppose that couples in one population (A) produce an average of four children by the age of twenty but that in a second one (B) this number is produced only by the age of thirty. Both populations will double in size each generation but when their relative sizes are tabulated by chronological time, we get the following results:

Years from Beginning	Relative Sizes of Populations A	B	Ratio A/B
0	1.0	1.0	1.0
10	1.4	1.3	1.1
20	2	1.6	1.2
30	2.8	2	1.4
40	4	2.5	1.6
50	5.7	3.2	1.8
60	8	4	2.0
70	11.3	5.0	2.3
80	16	6.4	2.5
90	22.6	8	2.8
100	32	10.1	3.2
110	45.3	12.7	3.6
120	64	16	4.0

Within 60 years population A becomes twice as large as B and, within another 60 years, it becomes four times as large. The ratio A/B itself undergoes an exponential increase! Even though the number of children born to different mothers may eventually prove to be the same (as in this example), it is the teen-age mothers, not those women who reproduce in their late twenties, who must bear the blame for explosive population increases.

A solid basis for a disaster in Ireland had been laid: an expanding population, a single food crop, small plots yielding annually a one year's food supply, and a high population density for both food crop and consumer. An additional factor contributing to the general instability of the entire system — like an additional card placed atop a card house — is the inability of potatoes to tolerate prolonged storage; a bumper crop of potatoes can be stored over winter, but it cannot be kept in storage year after year as insurance against the failure of future harvests.

The biological basis for a disaster in the 19th century Ireland has now been adequately described. To the biological facts already cited we can add an additional, political one: the British Corn Laws. Under these laws it was illegal to export or import grain from or to the British Isles; these laws were intended to protect the income of British farmers and to keep the nation self-sufficient.

In 1845 the house of cards collapsed. A fungal blight that had appeared only the year before in America struck the Irish potato crop. Not only was the harvest reduced but, in addition, seemingly sound potatoes that were shipped to market for sale rotted within a few days. The blight spread from plot to plot, from neighborhood to neighborhood, with tremendous speed. Because potatoes spoil during prolonged storage, there were no food reserves to be tapped; the Irish of the 1840's were entirely dependent year after year upon the annual potato harvest. Furthermore, food for the relief of the starving could not be imported until the Corn Laws, venerated by the British for five centuries, could be repealed. Disaster struck and continued to strike for several years. Potatoes normally saved for seed were eaten; only the eyes were kept for planting. From the eyes, however, abnormally small plants develop because the sprouting plant requires for its growth the food that is stored in the seed potato itself. The small plants that arise from eyes do not yield many potatoes at harvest time. The disaster in Ireland became a downward spiral from which the country has only recently recovered. Marriage patterns in Ireland have never been the same again. Today Irishmen do not marry until their fathers are ready to retire nor, apparently, are they willing to accept wives who are much younger than themselves. Contrast these modern marriage customs with the old, with the splitting and resplitting of family holdings to accommodate the offspring of teen-age marriages that preceded the disaster of the mid-19th century.

The events that transpired in Ireland before and during the potato famine illustrate the grossly uneven approach of populations toward equilibrium conditions. Had someone consciously planned to manipulate the Irish into a situation from which they could not escape without suffering severely, he could not have done better than to bid the population to follow the course it took of its own volition. And how about the United States today? Or, in broader terms, how about any of the highly industrialized Western nations?

To an extent that is difficult to admit, we have laid the groundwork for a similar catastrophe. Surplus foods no longer exist; they have long since been used to support — temporarily — the expanding populations of other nations. Fewer and fewer persons are engaged in agriculture; mechanization has made farming so efficient — superficially at least (see the following essay entitled "On the High Cost of Food") — that less than 1 person in 20 need farm in order to feed the remaining 19 people. The population continues to grow at an alarming rate. The ever-increasing numbers of persons require more and better highways and more and more housing. For the most part, land acquired to meet these needs is prime farmland because such land is easy and inexpensive to work. In the name of efficiency and economy, the farmland in large segments of the country is devoted to single crops, a procedure that encourages the spread of disease and insect pests; heroic — but not necessarily intelligent — measures are required to prop up this unstable arrangement.

Consequently, it appears that in the 20th, as in the 19th, century populations tend to erect shaky edifices that will crumble whenever outside stresses become sufficiently great. Although we do not have corn laws in the United States, we have in their stead an entire superstructure of distribution problems that accompany the urbanization of the population and the specialization of large areas of the country for specific crops. The entire complicated system runs because it has not yet been overburdened to the point of breakdown. This unhappy event, as in the case of the potato famine, may come with unchecked growth in population size.

On the High Cost of Food

Food has two costs: the price one pays at the supermarket and the energy spent in harvesting and getting it to the market. Customer complaints about the monetary cost of food are legion; at the risk of alienating shoppers, however, dollars and cents will not be discussed in this essay. Rather, the discussion will deal with the other cost, with the expenditure of calories needed to bring home one calorie of food. If more calories are expended on obtaining food than one recovers from the food obtained, the individual or population involved is living on capital and, unless the quantities involved are negligible, the system must eventually run down and stop. That is the point of this essay. I shall take the opportunity to point out once more, however, that the numbers of persons living today are not negligible.

The importance of the caloric cost of food is best illustrated by removing the consumer from civilization and its dollar distractions. Consider instead a shipwrecked sailor cast upon a lonely island where his only plentiful supply of food consists of numerous sleek fish that dart in and about the lagoon by which he has built his driftwood hut. The food is his for the catching; no checker awaits at the cash register nor game warden on the beach; he merely carves himself a sharp stick and dives in the water — lunging, swimming, diving, and spearing. And so, provided that he has a pan and a fire, it appears that he is well off indeed. Or is he? Appearances are sometimes deceiving; to see how our sailor will fare, we must total up the caloric costs.

A burst of violent physical activity lasting 10 seconds requires at least 6,000 cubic centimeters (6 liters) of oxygen. The sugar that is burned in combining with this much oxygen generates about 30,000 gram-calories (the chemist's calorie) or 30 kilogram-calories (the dietitian's calorie) of heat. During underwater activity, a person loses heat to the water about him. In 10 seconds of constant motion, a swimmer may bring a film of water covering the surface of his body and having a thickness of one-half centimeter from 70 F (water temperature) to 98.6 F (body temperature); the loss of heat to this film of water may amount to as much as 150 calories or more during 10 seconds of activity. Thus, anywhere from 175 to 200 calories may be expended in 10 seconds of spear fishing.

The Handbook of Chemistry and Physics tells me that 100 grams of fish supply a person with approximately 100 calories; some fish (oily ones) supply more and others less but the figure cited is a fair average. And so, to replace the calories spent in a 10-second period of underwater physical activity, our sailor must catch 175 to 200 grams of edible fish. If we estimate that about half of a whole fish is edible, we see that 350 to 400 grams or nearly a pound of fish must be caught to replace the expended energy during a 10-second foray.

To a nonswimmer, nonfisher the need to return to the beach at the end of a 10-second hunt with a one-pound fish is discouraging to contemplate. Nevertheless, to return with a smaller catch would be worse than relaxing in the shade with a painfully empty stomach. A search for less agile prey — abalones or other shell fish — would be a far better occupation than spear diving. Even when money is not involved, there are "expensive" and "inexpensive" foods.

The example used here to illustrate the cost of food in terms of calories is not as farfetched as the use of a shipwrecked sailor and a desert isle may imply. First, this aspect of the cost of food was confirmed by Jacques Cousteau and his co-workers during World War II. Even though they were conducting research on skin-diving techniques, they could not profitably supplement their wartime quota of meat by spearing fish; in the attempt they consistently lost more energy than they gained. Second, throughout the animal kingdom there are animals of one sort (predators) that devour others (prey). The prey species have a number of options open to them to avoid being eaten. They may develop retaliatory weapons for self-defense. They can rely on speed in order to escape by fleeing. They can develop protective shells. They may possess cryptic colors and shapes that make them hard to find. Some develop poisonous or other obnoxious characteristics and then advertise this fact by what are known as warning colorations. Some may *pretend* that they have developed such obnoxious or harmful characteristics by mimicking the warning colors of others.

Still another option has been utilized by certain animals that might otherwise be eaten as prey. These animals are generally energetic. They are strikingly colored but are neither obnoxious (they will be devoured readily if hand-fed to appropriate predators) nor pretending to be obnoxious (their color patterns do not mimic those of obnoxious models). The simplest explanation for such cases is the following: these animals (among them insects, fish, and mammals) are simply informing potential predators, "I am exceedingly fast and elusive; you will burn more energy in capturing me — assuming that you do — than I will yield in return." And the workings of natural selection are such that the advertisement works; predators make no serious effort to pursue these relatively defenseless but elusive creatures.

There is, therefore, a grand economy to getting food that serves not only to amuse us in reference to shipwrecked sailors but to enlighten us in reference to the "whys" and "why nots" of certain prey-predator relationships in the world about us.

How is man faring? More precisely, how are the agricultural activities of the United States faring? Quite good, it seems. The farm population of the United States in 1967 was 11,000,000 persons; in 1960 it had been 16,000,000. The total population of the United States in 1967 was 200,000,000; in 1960, 180,000,000. Thus, 1 person in 20 was providing food for the other 19 in 1967;

1 person in 11 was supplying food for the other 10 in 1960. This is one of the criteria by which efficiency may be measured.

Efficiency, as we have learned in this essay, has other measures; it is inefficient to spend more calories in the search for food than is obtained in a successful hunt. Such hunts are not really successful and, indeed, do not generally occur in the natural world. Under this criterion, how does the agriculture of the United States fare? The following table gives us a starting point for a discussion.

Total Harvest

Meat production*	16,000,000,000 kg	16×10^{12} cal
Lard	1,000,000,000 "	9 " "
Milk	30,000,000,000 "	20 " "
Eggs	70,000,000,000 eggs	6 " "
Potatoes	15,000,000,000 kg	15 " "
Peanuts	1,000,000,000 "	5 " "
Sugar	30,000,000,000 "	120 " "
	Total calories	191×10^{12}
		$= 19 \times 10^{13}$

*The bulk of the small grains (corn, wheat, etc.) is used in the production of meats and meat products and so these are not listed twice.

Partial Cost

Nonhighway gasoline consumption	12,000,000,000 kg	14×10^{13} cal

If nonhighway gasoline consumption is used primarily on the farms of America (and, despite the growing number of marinas, I believe it is), there is very nearly a 1:1 correspondence of calories burned to calories harvested (kilogram-calories have been used exclusively in the above table). To this partial cost of the harvest, however, we should add the gasoline used in the nationwide distribution of food and the electrical energy needed to produce the nitrogen- and phosphorus-containing fertilizers. We should also include the energy required for casting, first in the form of iron "pigs" and then into final molds, of the iron and steel that goes into the huge trucks that carry farm produce on the nation's highways and the farm equipment that makes the American farmer appear so efficient, at least superficially. When these extra caloric expenses are added, the cost of food in the United States is high indeed; much more energy is expended in obtaining our food than the food is worth after we get it.

"Better," I hear someone say, "that the calories for harvesting crops come from an oil can than from my wife's back." Very true! And, if mankind were restricted to numbers of individuals commensurate with his physical size and caloric requirements, energy from the gasoline tank would be a marvelous contribution to a civilized way of life. The sheer numbers of persons on earth, however, make it impossible to continue for long the deficit-spending that is characteristic of agriculture in industrialized nations and misleading to extol these Western-style procedures to other, less developed, more densely populated nations as a successful way of life.

No matter how the situation is rationalized, our food-getting procedures are at best only temporarily successful. The reckoning may not hit the present, middle-aged generation; it will in all likelihood hit our grandchildren. There may be an euphoria that envelops a skin diver who returns from an underwater hunt with a one- or two-pound fish on his spear; in the long run, however, he cannot live on euphoria. The highly touted agricultural system of the United States is at the moment euphoric; the euphoria will vanish when the true cost of food for feeding hundreds of millions or billions of persons becomes apparent.

On Population Control

During recent months I have met with 20 students for two hours each week. Our discussions have been largely devoted to procedures by which the size of a nation's population can be controlled.

These are intelligent students who understand the immensity of numbers such as "billion" and "trillion." They understand, too, the enormous demands that man makes on his surroundings, both in the sense of satisfying basic physiological needs and in the sense of meeting extraneous desires that consume energy and irreplaceable energy sources in wholesale quantities. They also understand that the world has reached a point of crisis; there are too many people making too many demands of a limited earth. There are far too many people who have not yet made demands to which (under the Golden Rule, if no other) they are entitled. In our discussions, these students were willing to assume that there are no major untapped resources for supporting much larger numbers of people on earth than now exist.

Despite the depressing constraints within which our discussions were carried out, no effective procedure for controlling population size emerged spontaneously. The contents of the early discussions were no different from those displayed in the Letters-to-the-Editor section of nearly any local newspaper. The tragedy of the commons was enacted within the classroom. If a solution to the population problem had depended upon the early discussions of this group of students, the problem would have remained unsolved and the population explosion would have continued unabated.

The population problem, however, cannot be left unsolved; consequently, the debates and dialogues continued. At last, there emerged three propositions that were operationally sound, that could be defended by rational arguments, but that were not defensible on existing moral or ethical grounds. These propositions and their meaning form the basis for this essay. The propositions are not new in the sense that they are assembled here for the first time. Nevertheless, they differ sufficiently from the usual starting points in discussions of population control so that they promise to lead to something new if pursued thoughtfully.

The three propositions are as follows:

1. Human life begins at the successful termination of a desired birth. Before its birth, the fetus possesses fetal, not human, life. Following birth, the newborn baby joins the human community and is entitled to

the same care, respect, and protection that is afforded every older member of the community.

2. The unborn fetus, from the moment that pregnancy is first diagnosed until it is born following willful decisions on the part of the mother, is part of the mother and exists at the pleasure of the mother – the mother alone; the father, relatives, physicians, hospital boards, and other groups of citizens in any guise whatsoever have not the slightest part in determining whether an unborn child will or will not be carried to term. In modern parlance, all persons (including the un-person, the fetus itself) other than the mother are irrelevant.

3. Society, through its government, can offer financial rewards on an annual basis to each woman of childbearing age (ages 12 through 47 for example) who has not borne a child during the past year. This reward is in payment for a valuable service rendered to society by these women. The precise amount of the reward can be adjusted according to each woman's age and to the size of the existing population.

What do these propositions accomplish? The first removes the danger that the legalization of abortion will lead in turn to the legalization of euthanasia or of any other type of legalized killing. Birth is an abrupt event; if it were not, birth dates and birth certificates would not exist. Following birth, according to the first proposition, human life is sacred in the sense that individual members of the human community are to be protected by the community at large. No one should be subjected needlessly to sustained physical pain or to worried anguish over what the future may hold for him at the hands of society. By the definition of human life given above, no living person need fear that his existence is threatened by legalized abortion. Furthermore, because fetuses are isolated from one another and from the outside world, one fetus cannot possibly suffer mental anguish as a result of another's abortion.

Several years ago an attempt was made to liberalize the abortion laws of New York State. The attempt failed largely because of a last-minute opposing address delivered by an assemblyman wearing leg braces. The theme of this speech was misleading. The speaker referred to himself as a cripple and suggested that had this fact been known before his birth he would not have been permitted to live. The suggestion is faulty in the trivial sense that his deformity was caused by polio contracted as a child, not by a congenital or prenatal event. More important, however, is the error in his belief that because he is happy that he now exists, he would be unhappy if he did *not* exist. Persons who hold this view simply have not grasped the meaning of nonexistence. The crippled assemblyman is a welcome member of the human community *because* he does exist; if he did not exist, no one would be concerned in the slightest, himself least of all.

The second proposition, one that will be welcomed by the more militant feminists among us, places the fate of the fetus in the hands of its mother. It is growing in the mother and is a part of the mother. It has not attained membership in the human community. The proposition says that the removal of an unwanted fetus is no different than the removal of any other unwanted growth of living tissue. The removal should not be contingent upon extensive and time-consuming (footdragging, in many instances) reviews by committees and panels (usually of men, not women) who in the final analysis are really not concerned with the problem in any logical way. Furthermore, the role of these panels and committees, whatever it may now be, has been eliminated by the first proposition.

The third proposition recognizes that in a dangerously expanding population a woman who refrains from childbirth has performed a service for the community and should be rewarded in return. It also recognizes that the reward should be made on an annual basis so that no woman is asked to make an irrevocable lifelong decision. A woman who remains childless throughout her lifetime, however, would receive about 35 annual payments — presumably a sizable amount of money in all. Finally, the proposition recognizes that the value of the service rendered by the temporarily childless woman increases as population numbers approach some predetermined goal; consequently, as the population grows, the reward is increased. In short, as the population grows, the ante is raised so that a greater proportion of all women will find it worth their while to remain childless each year.

The propositions outlined here cannot be defended, as far as I can see, on present-day moral or ethical grounds. On the other hand, I am not convinced that currently accepted moral and ethical arguments (which may prove fallacious, incidentally, in the light of recent advances in developmental biology) would be as appealing as they now are if the population were one eking out a marginal existence perpetually on the brink of mass starvation. The matter of the care and respect accorded to human life has been solved in the first proposition *by definition*. In what state, however, is the dignity of human life in a population where one-half or more of all children die before the age of five and where a high proportion of adults succumb to loathesome diseases that are augmented by semistarvation diets? The prevention of a birth, by abortion if necessary, is preferable to the loss of a young child through starvation and at a price of acute physical suffering.

The preceding proposals remain unrealistic unless several preconditions are met. To a large extent these preconditions are already in operation within the United States. The first of these is that education in sexual matters be extended to all persons before physical maturity; such education does exist today although it exists not without considerable opposition from some segments of the population. Second, birth control materials should be at the unhindered disposal of all who need them; this condition merely acknowledges that the individual's private behavior is not a legitimate concern of the public. On the other hand, the possible consequences of that behavior — the birth of new persons — are not entirely a matter of public indifference. Third, should an unwanted or unexpected pregnancy occur, the woman — under the provisions of the second proposition — can have it terminated at any moment and at a minimal fee (if any) under decent clinical conditions; no hurdles other perhaps than a 72-hour waiting period (such as many states require of applicants for marriage licenses) are to be placed in the way of the woman who requests an abortion.

Still another precondition must be met before the propositions of this essay make a great deal of sense. A guaranteed annual income that meets the basic needs for food, clothing, and shelter for everyone must be in effect. Fortunately, the idea of a guaranteed annual income has been suggested by President Nixon's administration as an alternative to present welfare procedures. Presumably, the system, if put into effect, would provide for increases in basic income to meet the extra financial burden of dependent children. The propositions we have advanced would provide an annual bonus of comparable size (perhaps larger, in order to emphasize the need for population control) for not having a child, a bonus that could be used by the woman not to meet greater living expenses but to supplement her or her husband's income.

The issue of population control must be joined. For too long persons have ignored the inexorable increase in their numbers and have hoped that food technologists would emerge with a solution for providing food for all. The hope that food production can solve the population problem is demonstrably futile. That the sheer numbers of human beings are straining the material bases for life is becoming more obvious each day. This is the background against which a discussion of population control must be held. If a background of illusory optimism is substituted, the age-old ethical and moral issues rise up to entangle the discussion and to render it inconclusive. This essay, whatever else may be said about it, has maintained the stark facts of uncontrolled population growth as a backdrop. It has presented a series of propositions that would lead to population control without coercion. It has provided for the protection of each individual against capricious acts by the society in which he lives. And it has provided a basis for discussions that might lead to a rational and acceptable system of population control even though that system may come to differ considerably from the one proposed here.

The Tortuous Road to
Population Control*

William F. Bundy

So quietly that it attracted only minor attention in the press, the Senate has just passed a bill to authorize a major long-term program of birth-control assistance and research in this country. . . .A simple action, perhaps, and to many people long overdue. Yet nothing could more vividly symbolize the revolution of the last decade in American thinking and practice on this literally vital issue. . . .No such change was foreseen in 1960, when I wrote a chapter on this subject for President Eisenhower's Commission on National Goals. Today, by contrast, the safest prediction one could make for the United States in 1980 is that the downward trend [in growth rate] will have continued. We may even be coming fairly close to a reproduction rate of two children per couple as a national average. So it seems to me that arguments about whether the United States have a population problem are beside the point. The American people have made up their minds. . . . All this is a healthy trend, even if its motivation is sometimes oversimplified. For example, population growth is not really the major factor in the physical pollution of the environment: Growth per person in living standards, and thus in raw materials consumed and waste produced, is far more important. . . . The same is true to a significant extent of the problem of world food balance: changes in eating habits — mostly, of course, for the good — are for many key items at least as important as changes in the number of mouths. The population·problem is neither a major villain nor a reasonable excuse in either area. . . . One can sum up in a basic proposition that goes far beyond family planning or any existing program: Only if there is a drastic change in the way mankind as a whole looks at family size — and then only if this change in thinking is maintained for a number of decades — can the growth of human population be brought to levels in line with man's capacity to handle change. . . . Mr. Nixon's new "national goals research staff" told us the obvious

*Condensed from an article entitled "Population at Home and Abroad" that appeared in the International Edition of NEWSWEEK MAGAZINE on August 10, 1970. Copyright Newsweek, Inc., August 10, 1970.

the other day: that the problem in the United States is not food, total space, or even energy supply, but rather the ever growing choice of a few congested areas in which to live. . . . I think the change in attitudes [toward crowdedness] can come without force or major crisis. For come it surely must, and soon. Robert Ardrey wrote last winter that all animals save man had learned to curtail their population at a point of strain long short of exhaustion of key resources. I doubt if man, faced with the choice for the first time, will prove an exception.

● ● ●

Personal Commentary

Professor Bundy ends on what I regard as an unjustifiably optimistic note. Mouths to be fed already exceed, in my estimation, the number that can be fed Western-style – the style most sought after. Persons longing for the physical comforts enjoyed by citizens of developed nations already exceed the number that can be granted such comforts.

In his encyclical, Pope Paul VI makes the following point: "Children are really the supreme gift of marriage and *contribute very substantially to the welfare of their parents*" (emphasis mine). In a country that lacks an effective social security system, a man who limits the number of his children acts irrationally. How many countries, however, have workable social security systems? Why then should we expect any "drastic change in the way mankind as a whole looks at family size?"

The preceding comment can be used as a pretext for discussing a similar point made by Wallich: "With some reason. . . .India and other developing countries have been internationally lectured on the virtues of birth control. . . . Now, this good advice is being sent to a new address – the U. S. . . . The charge of overpopulation could hardly have been addressed to a more inappropriate country."* In continuing the comment made in response to Professor Bundy, I would ask if it is not necessary for some nation to take the lead in setting new goals affecting the quality of life. What better nation, therefore, than the United States? Only when the United States and other developed nations have accepted such goals can they help reshape the social fabric of "underdeveloped" nations in a way that makes birth control a rational act for the individual citizens of those nations. There are certain undertakings (of which population control is one) for which mankind must recognize its fundamental unity.

*Henry C. Wallich, "On Population Growth," *Newsweek*, June 29, 1970. Subsequent issues contain many letters in response to this essay, as well as Wallich's reply.

On Decision Making

Decisions, decisions! Throughout each day countless small decisions must be made. Fortunately for most of us the consequences of one choice are, as a rule, not materially different from those of another. Do I go to the library and risk missing a phone call, or do I work in my office until the call has been received? Do I walk to work, or do I take the car? Dozens of matters such as these arise; they are settled with scarcely a conscious thought because they are so unimportant: life continues unchanged whether the phone call precedes or follows a visit to the library; the ten-minute walk leaves the rest of the day unaffected.

Not all decisions are trivial ones. A traveler in a strange city must cease sightseeing in time to catch his airport limousine. As a rule he arrives at the air terminal early because failure to show might cost him his flight reservation. An offer from another university causes the average professor days or weeks of agonizing stock-taking: salary, housing, friends, education for the children, opportunities for advancement, provisions for retirement, effectiveness as an educator, and other considerations must be weighed in some rational manner. A decision that will affect the remainder of a person's life together with those of his wife and children is ordinarily not taken lightly.

Decision making has been reduced to a formal art by mathematicians; it falls within a branch of mathematics known as *game theory*. The ultimate outcomes of alternative decisions are listed together with an estimated likelihood for each. Each outcome is assigned a value according to its effect on the "player." The products of the likelihoods and values offer a basis for identifying the right decision. It was, for example, the importance (that is, high value) of catching the plane in the earlier example that got the traveler to the air terminal in ample time. The pleasure of spending Thanksgiving with a reunited family induces many onto the highway for holiday travel, a decision that for some proves to be fatal.

Decision making by means of game theory encompasses decisions involving many persons where the numerous outcomes possess different values for each player. The correct decision, the one presumably arrived at by the democratic process, is the one that leads to the maximum combined value for all concerned. Herein lies the germ for decisions bearing on the establishment of acceptable levels of air and water pollution, on the appropriate use of farmland by an expanding population, on the regulation of the use of the internal combustion engine, on the prohibition of certain food additives, and on a multitude of other problems that involve sharp conflicts of interest between segments of society.

Like all mathematical procedures, game theory is useful in the process of decision making only to the extent that its underlying assumptions are met by conditions existing in the real world. Two of these assumptions are especially important for political decisions. Each player is assumed to have full knowledge of the game in extensive form; not only is he assumed to know the rules of the game in full detail but also the payoff functions of the other players. Furthermore, each player is assumed to be *rational* in the sense that given two alternatives, he will always choose the one he prefers, that is, the one with the larger utility.

The above paragraph is couched largely in the formal words of mathematics.* Nevertheless, the implications of these words for society shine through remarkably well. To have *full knowledge* of the game requires, in society, massive educational programs. The book you are now reading and the many others that deal with technology and society exist to help each player gain his full knowledge of the game of life. To have full knowledge in *extensive form* requires that the boundary delineating the area within which a decision must be reached not be drawn too soon; if it should be, the "tragedy of the commons" is reenacted once more. Secretary McNamara, according to critics of his Vietnam policies, failed to extend his decision-making diagrams far enough; he misjudged both the payoff functions and the rationality of all players (including that of his own people). The *full* implication of modern technological changes must be understood before they are approved; in large measure this means an understanding of the finiteness of the earth itself and of its resources. The earth can no longer be left outside the boundary that designates the elements that enter into certain decision-making processes. Events transpiring in Northern cities, for example, cannot be omitted from analyses of Southern agricultural progress.

Vested interests have ruses by which they bias decisions in their favor. Some of these involve unilateral acts performed early without fanfare and designed to present opponents with a fait accompli. Millions of dollars may be spent on preliminary site preparations for a construction project in the hope that no rational person would then oppose the project as a whole. Fortunately, persons sometimes do; a newly erected, 30-story apartment house was recently demolished in Brussels because it failed to meet zoning requirements.

In some instances the outer boundary is deliberately drawn so that the "extensive form" requirement of game theory is circumvented. This is the basis of most advertisements urging this or that course of action; each advertisement emphasizes the immediate impact of a given act with no mention of its ramifications nor of the larger consequences that would follow its adoption. Here lies much of the conflict between ecologists and the jetport builders, the highway men, and the spokesmen for the pesticide industry.

---

*For example, see R. Duncan Luce and Howard Raiffa, *Games and Decisions* (New York: John Wiley & Sons, Inc., 1957).

The citizens of a country represent the players in the game of decision making. Collectively, they form a diffuse element in the decision-making process; the majority of them, in fact, are as yet unborn. Citizens speak through their governmental representatives at the local, state, and federal levels. As members of the family of man, their representatives meet in international forums. These representatives — civil servants, in the best sense of this phrase — have the authority to impose regulations binding on all, on the individual citizen as well as on the larger, more powerful corporate body. At times, a representative can anticipate the true interests of his constituency and act counter to his expressed mandate; he cannot, however, often act counter to the immediate desires of his electorate, for if he does he will not be returned to office.

People do what people want to do. Many persons recite this refrain as if it alone absolves them of further responsibility; having recited their piece, they step back to join other interested spectators. Others, however, believe that education — the promotion of full knowledge in extensive form — will alter what people *want* to do, that education will lead people to a better understanding of the strengths and frailities of their good earth. These other persons are the gadflies of much of modern industry. They are skeptical about much of modern technology. They are largely nonconformist. And, most dangerous of all in the eyes of their opponents, they strive within the political arena to improve the decision-making role of the ordinary citizen by promoting full knowledge, by pushing back the boundary that encloses each problem for which a solution is sought, and by pointing out where the citizen's greatest utility really lies.*

*The opponents, of course, have countermeasures that are temporarily effective. According to a news release dated November 13, 1970, the executive branch of the federal government has decided to withhold environmental impact studies from the public *until the decisions they influence have been made and announced.* This unfortunate decision will remain in effect only as long as large numbers of citizens tolerate it; an incensed public can have it revoked. (Indeed, during the production of this book the order was rescinded.)

SECTION TWO

The Environment

Introduction

The outstanding feature of today's environment is the speed with which it is deteriorating. Not that environmental problems are new, for they are not. They are simply enlarging and threatening to become uncontrollable. Man's effort to support vast numbers of persons on an ever-shrinking agricultural base is, in the absence of population control, intrinsically futile. Much of the deterioration of his environment stems from frenzied, stopgap measures that shore up, postpone, but do not cure.

What constitutes a deteriorating environment? Ordinarily, sane persons would agree that the wholesale distribution of poisons throughout the countryside — the dispersion of lead and carbon monoxide by millions of automobiles and trucks, the spreading of highly toxic, long-lasting pesticides of all sorts on thousands and thousands of square miles of farmlands and forests, and the production of tons of airborne sulfurous and nitrogenous fumes by industry — is a dangerous and undesirable act. These, however, are not ordinary times; in our haste to keep food production abreast of population numbers, we can scarcely avoid poisoning those we hope to save. How better can one illustrate the utter absurdity of today's way of life?

Human wastes constitute a large portion of the human environment. Their disposal has plagued man from earliest time. When people were few in number and space was plentiful, a population simply moved away from its rubbish heaps. When space became limited, wastes were thrown into rivers to be washed away to the sea. Few societies, apparently, have understood that organic wastes (urine, feces, and garbage) represent but one link in the cycle of nutrients through the skein of life. Few persons of any society understand the significance of the words "cycle" and "recycle." Wastes washed from the land exhaust the nutrients of the land. If the tale is true, the ancient rural Chinese were a notable exception of their appreciation of food cycles; to leave a dinner party without first depositing one's "night soil" at his host's home was supposedly a serious breach of etiquette. The thoughtless guest who violated this custom not only ate his host's food but also deprived him of the nutrients needed for growing an equivalent crop of garden vegetables.

The selections chosen to introduce the problem of environmental pollution illustrate the snail's pace with which society responds to its challenges. John Evelyn described the dense smoke of London and its causes in 1661. One hundred years later, when Evelyn's pamphlet was reissued, the list of offending industries had grown and the effects of the "smoake" had worsened. Three centuries later the smog was still worse. The sources of smoke and smog had continued to multiply; in the early 1960's Londoners saved themselves during the worst smogs by hiding indoors or by wearing face masks. Finally, in 1970, news from London of a smogless November was published in many newspapers abroad. Effective measures against air pollution are apparently not impossible to enforce; it only requires centuries for citizens to act.

Mr. Wylie's many-faceted account of pollution in and about Britain makes both fascinating and charming reading. His description of the surging of the sludge back and forth in the mouth of the Thames River reminds me of the reportedly similar surging of invisible radioactive wastes from the nuclear submarines based at the Firth of Forth. The hot water discharged by industrial plants into lakes also refuses to become diluted and to dissipate as simple calculations based on relative volumes suggest they should.

The selection from Steinbeck's *The Grapes of Wrath* illustrates a deterioration of the environment not described any further in this collection of essays — the deterioration of the land that follows ill-advised agricultural practices. In 1938, the year preceding Steinbeck's novel, the Department of Agriculture reported that over half of the land area of the United States had suffered moderate or severe erosion, where "moderate" was defined as the loss of 25 per cent to 75 per cent of the original topsoil and "severe," as more than 75 per cent. It is ironic that the refugees from the nation's Dust Bowl, together with the defense workers who streamed into Southern California during World War II, spelled the doom of the fertile farm valleys surrounding Los Angeles. Presumably the topsoil of these valleys is still there, uneroded but buried irretrievably under thousands of city streets and parking lots. Meanwhile, on these streets are found millions of private automobiles, the only means of transportation serving widely dispersed families. It is this horde of cars that causes the notorious Los Angeles smog, which in spreading over the surrounding countryside has virtually eliminated pine forests on mountain ranges more than 100 miles away from the city itself.

Fumifugium; Or Smoake of London Dissipated

John Evelyn

Preface to the Edition of the Tract

Reprinted for B. White, in Fleet Street, 1772

The established reputation of Mr. Evelyn's writings would have prevented the Editor of this very scarce Tract from adding any thing himself, had not time made some alterations that appear worthy of notice.

Our Author expresses himself with proper warmth and indignation against the absurd policy of allowing brewers, dyers, soap-boilers, and lime-burners, to intermix their noisome works against the dwelling-houses in the city and suburbs; but since his time we have a great increase of glass-houses, foundries, and sugar-bakers, to add to the black catalogue; at the head of which must be placed the fire-engines of the water-works at London Bridge and York Buildings, which (whilst they are working) leave the astonished spectator at a loss to determine whether they do not tend to poison and destroy some of the inhabitants by their smoke and stench than they supply with their water. Our author also complains that the gardens about London would no longer bear fruit, and gives instances of orchards in Barbican and the Strand that were observed to have a good crop the year in which Newcastle was besieged (1644), because but a small quantity of coals were brought to London that year; by this we may observe how much the evil has increased since the time this treatise was written. It would now puzzle the most skillful gardener to keep fruit trees alive in these places: the complaint at this time would be, not that the trees were without fruit, but that they would not bear even leaves.

Although the proposal of turning all the noxious trades at once out of town may be thought impracticable, as being inconsistent with the general liberty of the subject; yet certainly some very beneficial regulations lie within the power of the present public-spirited and active magistrates, to whom, with deference, the editor submits the following hints.

Till more effectual methods can take place, it would be of great service to oblige all those trades, who make use of large fires, to carry their chimneys much higher into the air than they are at present; this expedient would frequently help to convey the smoke away above the buildings, and in a great measure disperse it into distant parts, without its falling on the houses below.

Workmen should be consulted, and encouraged to make experiments, whether particular construction of the chimneys would not assist in conveying off the smoke, and not sending it higher into the air before it is dispersed.

A method of charring sea-coal, so as to divert it of its smoke and yet leave it serviceable for many purposes, should be made the object of a very strict enquiry: and premiums should be given to those that were successful in it. Proper indulgences might be made of such sugar, glass, brewhouses, etc., as should be built at a desired distance from town: and the building of more within the city and suburbs prevented by law. This method vigorously persisted in, would in time remove them all.

The discernment and good sense of the present times are loudly called on to abolish the strange custom of laying the dead to rot amongst the living, by burying in churches and church-yards within town: this practice has not escaped our author's censure: and foreigners have often exposed absurdity of the proceeding. But it seems to be left particularly to the magistracy and citizens of London, to set an example to the rest of this kingdom and to Europe, by removing a nuisance which ignorance and superstition have entailed on us hitherto; and which, amongst those that are not well acquainted with our religion, brings disgrace on Christianity itself. It will be a work of little shew or ostentation, but the benefits from it will be very extensive and considerable: in both respects it recommends itself in a particular manner to an opulend and free people.

To confirm what our author has urged against the air in London, the reader is desired to take a view of the Bills of Mortality, and the calculations made from them; and he will find that there is a waste of near 10 thousand people, who are drawn every year from the country to supply the room of those that London destroys beyond what it raises. Indeed the supply that the town furnishes towards keeping up its own inhabitants appeared so very small to the ablest calculator and most rational enquirer (Corbyn Morris) into this subject, that he owns he was afraid to publish the result.

But, without the use of calculations, it is evident to every one who looks on the yearly Bill of Mortality, that near half the children that are born and bred in London die under two years of age. Some have attributed this amazing destruction to luxury and spirituous liquors: – these, no doubt, are powerful assistants: but the constant and unremitting poison is communicated by the foul air, which, as the town still grows larger, has made regular and steady advances in its fatal influence.

The ancient Greeks and Romans, even in their greatest state of refinement, were reconciled by habit to the custom of exposing and destroying young children, when parents did not choose to support them: the same practice is familiar among Chinese at this day. We shudder and are shocked by the barbarity of it, but at the same time are accustomed to read with great composure of the deaths of thousands of infants suffocated every year by smoke and stenches, which good policy might in a great measure remove.

Our author, who had been instrumental in restoring Charles to his throne, was unfortunate in recommending a work of such consequences to so negligent and dissipated a patron. The editor is encouraged by a more promising appearance of success. He has seen with pleasure many improvements of great importance to the elegance and welfare of this city undertaken and completed in a short time, when Magistrates of less public spirit and perseverance than are present, would have pronounced them to have been impracticable. London, March 16th, 1772.

To the Kings Most Sacred Majesty,

Sir,

It was one day, as I was Walking in Your MAJESTIES Palace at WHITE-HALL, (where I have sometimes the honour to refresh myself with the

Sight of Your Illustrious Presence, which is the Joy of Your Peoples hearts) that a presumptuous Smoake issuing from one or two Tunnels neer Northumberland-house, and not far from Scotland-yard, did so invade the Court; that all the Rooms, Galleries, and Places about it were filled and infested with it; and that to such a degree, as Men could hardly discern one another for the Clowd, and none could support, without manifest Inconveniency. It was not this which did first suggest to me what I had long since conceived against this pernicious Accident, upon frequent observation; But it was this alone, and the trouble that it must needs procure to Your Sacred Majesty, as well as hazzard to Your Health, which kindled this Indignation of mine, against it, and was the occasion of what it has produced in these Papers.

Your Majesty, who is a Lover of noble Buildings, Gardens, Pictures, and all Royal Magnificences, must needs desire to be freed from this prodigious annoyance; and, which is so great an Enemy to their Lustre and Beauty, that where it once enters there can nothing remain long in its native Splendor and Perfection: Nor must I here forget that Illustrious and divine Princesse, Your Majesties only Sister, the now Dutchesse of Orleans, who at her Highnesse late being in this City, did in my hearing, complain of the Effects of this Smoake both in her Breast and Lungs, whilst She was in Your Majesties Palace. I cannot but greatly apprehend, that Your Majesty (who has been so long accustomed to the excellent Aer of other Countries) may be as much offended at it, in that regard also; especially since the Evil is so Epidemicall; indangering as well the Health of Your Subjects, as it sullies the Glory of this Your Imperial Seat.

Sir, I prepare in this short Discourse, an expedient how this pernicious Nuisance may be reformed; and offer at another also, by which the Aer may not only be freed from the present Inconveniency; but (that removed) to render not only Your Majesties Palace, but the whole City likewise, one of the sweetest, and most delicious Habitations in the World; and this, with little or no expence; but by improving those Plantations which Your Majesty so laudibly affects, in the moyst, depressed and Marshy Grounds about the Town to the Culture and Production of such things, as upon every gentle emission through the Aer, should so perfume the adjacent places with their breath; as if, by a certain charm, or innocent Magick, they were transferred to that part of Arabia, which is therefore styled the Happy, because it is amongst the Gums and precious Spices. Those who take notice of the Scent of the Orange flowers from the Rivage of Genoa, and St. Pietro dell' Arena; the Blossomes of the Rosemary from the Coasts of Spain many Leagues off at Sea; or the manifest and odoriferous wafts which flow from Fontenay and Vaugirard, even to Paris, in the season of Roses, with the contrary Effects of those less pleasing Smells from other accidents, will easily consent to what I suggest: And, I am able to enumerate a Catalogue of native Plants, and such as are familiar to our Country and Clime, whose redolent and agreeable Emissions would even ravish our senses, as well as perfectly improve and meliorate the Aer about London; and that, without the least prejudice to the Owners and Proprietors of the Land to be employed about it. But because I have treated of this more at large in another curious and noble subject, which I am preparing to present to Your Majesty, as God shall afford me Leisure to finish it, and that I give a Touch of it in this Discourse, I will enlarge by Addresses no farther, then to beg pardon for this Presumption of,

Sir,

Your Majesties ever Loyal, Most obedient

Subject and Servant,

J. EVELYN

To the Reader

I have little here to add to implore thy good opinion and approbation, after I have submitted this Essay to his Sacred Majesty: But as it is of universal benefit that I propound it; so I expect a civil entertainment and reception. I have, I confesse, been frequently displeased at the small advance and improvement of Publick Works in this Nation, wherein it seems to be much inferiour to the Countries and Kingdomes which are round about it; especially, during these late years of our sad Confusions. But now that God has miraculously restored to us our Prince, a Prince of so magnanimous and publick a Spirit, we may promise ourselves not only a recovery of our former Splendor; but also whatever any of our Neighbours enjoy of more universal benefit, for Health or Ornament: In summe, whatever may do honour to a Nation so perfectly capable of all advantages.

It is in order to this, that I have presumed to offer these few Proposals for the meliorating and refining the Aer of London: being extremely amazed, that where there is so great an affluence of all things which may render the People of this vast City, the most happy upon Earth, the sordid and accursed Avarice of some few particular Persons, should be suffered to prejudice the health and felicity of so many: That any Profit (besides what is of absolute necessity) should render men regardless of what chiefly imports them, when it may be purchased upon so easie conditions, and with so great advantages: For it is not happiness to possesse Gold, but to enjoy the Effects of it, and to know how to live cheerfully and in health, Non est vivere, sed valere vita. That men whose very Being is Aer, should not breath it freely when they may; but (as that Tyrant used his Vassals) condemn themselves to this misery & fumo prafocari, is strange stupidity: yet thus we see them walk and converse in London, pursued and haunted by that infernal Smoake, and the funest accidents which accompany it wheresoever they retire.

That this Glorious and Antient City, which from Wood might be rendred Brick (like another Rome) from Brick made Stone and Marble; which commands the Proud Ocean to the Indies, and reaches the farthest Antipodes, should wrap her stately head in Clowds of Smoake and Sulphur, so full of Stink and Darknesse, I deplore with just Indignation. That the Buildings should be composed of such a Congestion of mishapen and extravagant Houses; That the Streets should be so narrow and incommodious in the very Center, and busiest places of Intercourse; That there should be so ill and uneasie a form of Paving under foot, so troublesome and malicious a disposure of the Spouts and Gutters overhead, are particulars worthy of Reproof and Reformation; because it is hereby rendered a Labyrinth in its principal passages, and a continual wet day after the storm is over. Add to this the Deformity of so frequent Wharfes and Magazines of Wood, Coale, Boards, and other coarse Materials, most of them imploying the Places of the Noblest aspect for the situation of Palaces towards the goodly River, when they might with far lesse Disgrace, be removed to the Bank-side, and afterwards disposed with as much facility where the Consumption of these Commodities lyes; a Key in the mean time so contrived on London-side, as might render it lesse sensible of the Reciprocation of the Waters, for Use and Health infinitely superiour to what it now enjoys. These are the Desiderata which this great City labours under, and which we so much deplore. But I see the Dawning of a brighter day approach; We have a Prince who is Resolved to be a Father to his Country; and a Parliament whose Decrees and Resentments take their Impression from his Majesties great Genius, which studies only the Publick Good. It is from them therefore, that we Augure our future happinesse; since there is nothing which will so much perpetuate their Memories, or more justly merit it. Medails and Inscriptions have heretofore preserved the Fame of lesse Publick Benefits, and for the Repairing of a Dilapidated Bridge, a decaid Aquaeduct, the Paving of a Way, or draining a foggy Marsh, their Elogies and Reverses have out-lasted the Marbles, and been transmitted to future Ages, after so many thousand Revolutions: But this is the least of that which we Decree to our August CHARLES, and which is due to his Illustrious Senators; because they will live in our Hearts, and in our Records, which are more permanent and lasting.

Farewell.

May, 1661

Part I

It is not without some considerable Analogy, that sundry of the Philosophers have named the Aer the Vehicle of the Soul, as well as that of the Earth, and this frail Vessell of ours which contains it; since we all of us finde the benefit which we derive from it not onely for the necessity of common Respiration and functions of the Organs; but likewise for the use of the Spirits and Primigene Humors, which doe most neerly approach that Divine particle. But we shall not need to insist or refine much on this sublime Subject; and, perhaps, it might scandalize scrupulous Persons to pursue to the height it may possibly reach (as Diogenes and Anaximenes were wont to Deifie it) after we are past the AEtherial, which is certain Aer of Plato's denomination, as well as that of the lesse pure, more turbulent and dense, which, for the most part, we live and breath in, and which comes here to be examined as it relates to the design in hand, the City of London, and the environs about it.

It would doubtlesse be esteemed for a strange and extravagant Paradox, that one should affirme, that the Aer itself is many times a potent and great disposer to Rebellion; and that Insulary people, and indeed, most of the Septentrion Tracts, where this Medium is grosse and heavy, are extremely versatile and obnoxious to change both in Religious and Secular Affaires: Plant the Foote of your Compasses on the very Pole, and extend the other limb to 50 degrees of Latitude: bring it about 'till it describe the Circle, and then reade the Histories of those Nations inclusively, and make the Calculation. It must be confessed, that the Aer of those Climates, is not so pure and Desecate as those which are neerer the Tropicks, where the Continent is lesse ragged, and the Weather more constant and steady, as well as the Inclination and Temper of the Inhabitants.

But it is not here that I pretend to speculate upon these Causes, or nicely to examine the Discourses of the Stoicks and Peripateticks, whether the Aer be in itself generally cold, humid, warm, or exactly tempered so as best conduces to a materiall principle, of which it is accounted one of the four; because they are altogether Physicall notions, and do not come under our cognisance as a pure and sincere Element; but as it is particularly inquinated, infected, participating of the various Accidents, and informed by extrinsecal Causes, which render it noxious to the Inhabitants, who derive and make use of it for Life. Neverthelesse, for distinction sake, we may yet be allowed to repute some Aers pure, comparatively, viz. That which is cleare, open, sweetely ventilated and put into motion with gentle gales and breezes; not too sharp, but of a temperate constitution. In a word, That we pronounce for good and pure Aer, which heats not to sweat and faintnesse; nor cooles to rigidnesse and trembling; nor dries to wrinkles and hardnesse; nor moystens to resolution and over-much softnesse. The more hot promotes indeede the Witt, but is weak and trifling; and therefore Hippocrates speaks the Asiatique people Imbelles and Effeminate, though of a more artificiall and ingenious Spirit: If over cold and keen, it too much abates the heat, but renders the body robust and hardy; as those who are born under the Nothern Beare, are more fierce and stupid, caused by a certain internal Antiperistasis and universal Impulsion. The drier Aer is generally the more salutary and healthy, so it be not too sweltery, and infested with heat or fuliginous vapours, which is by no means a friend to health and Longaevity, as Avïcen notes of the AEthiops, who seldom arrived to any considerable old age. As much to be reproved is the moyst, viz. that which is over-mixed with aquous exhalations, equaly pernicious and susceptible of putrefaction; notwithstanding does it oftner produce faire and tender skins, and some last a long while in it; but commonly not so healthy, as in Aer which is more dry. But the impure and Uliginous, as that which proceedes from stagnated places, is, of all other, the most vile and Pestilent.

Now, that through all these diversitites of Aer, Mores Hominum do Corporis temperamentum sequi, is for the greater part so true an observation, that a Volume of Instances might be produced, if the Common notices did not sufficiently confirm it, even to a Proverb. The Aer on which we continually prey, perpetually inspiring matter to the Animall and Vitall Spirits, by which they become more or lesse obfuscated, clowed and rendered obnoxious; and therefore that Prince of Physitians Hippocrates, wittily calls a sincere and pure Aer, The Internunce and Interpreter of Prudence. The celestiall influences being so much retarded or assisted, and improved through this omnipresent, and, as it were, universal Medium: For, though the Aer in its simple substance cannot be vitiated; yet, in its prime qualities, it suffers these infinite mutations, both from superiour and inferiour Causes, so as its accidentall effects become almost innumerable;

Let it be farther considered, what is most evident, That the Body feeds upon Meats commonly, but at certain periods and stated times be it twice a day or oftener; whereas, upon the Aer, or what accompanies it (est enim in ipso Aere occultus vitae cibus) it is alwayes preying, sleeping or waking; and therefore, doubtlesse, the election of this constant and assiduous Food, should something concerne us, I affirme, more than even the very Meat we eat, whereof so little and indifferent nourishes and satisfies the most temperate and best Educated persons. Besides, Aer that is corrupt insinuates itself into the vital parts immediately; whereas the meats which we take, though never so ill conditioned, require time for the concoction, by which its effects are greatly mitigated; whereas the other, passing so speadily to the Lungs, and virtually to the Heart itself, is derived and communicated over the whole masse: In a word, as the Lucid and noble Aer clarifies the Blood subtilizes and excites it, cheering the Spirits and promoting digestion; so the dark and grosse (on the contrary) perturbs the Body, prohibits necessary Transpiration for the resolution and dissipation of ill Vapours, even to disturbance of the very Rational faculties, which the purer Aer does so far illuminate, as to have rendered some Men healthy and wise, even to Miracle. And therefore the Empoysoning of Aer, was ever esteemed no lesse fatall than the poysoning of Water or Meate itself, and forborn even amongst Barbarians, since (as is said) such Infections become more apt to insinuate themselves and betray the very Spirits, to which they have so neer a cognation. Some Aers we know are held to be Alexipharmac, and even deleterious to Poyson itself, as 'tis reported of that of Ireland: In some we finde Carcasses will hardly putrifie, in others again rot and fall to pieces immediately.

From these, or the like considerations, therefore, it might well proceed, that Vitruvius, and the rest who follow that Master-Builder, mention it as a Principle, for the accomplishment of their Architect, that being skilful in the Art of Physick, among other Observations, he sedulously examine the Aer and Situation of the places where he designs to build, the Inclinations of the Heavens, and the Climats; Sine his enim rationibus nulla salubris habitatio fieri potest: there is no dwelling can be safe or healthy without it. 'Tis true, he does likewise adde Water also, which is but a kinde of condensed Aer; though he might have observed, that Element to be seldome bad, where the other is goode; omitting onely some peculiar Fountains and Mineral waters, which are percolated through Mines and Metalique Earths, less frequent, and very rarely to be encountered.

Now whether those who were the Antient Founders of our goodly Metropolis, had considered these particulars (though long before Vitruvius) I can no waies doubt or make question of; since having respect to the nobleness of the situation of London, we shall every way finde it to have been consulted with all imaginable Advantages, not onely in relation to Profit, but to Health and Pleasure; and that, if there be any thing which seems to impeach the two last Transcendencies, it will be found to be but something Extrinsecal and Accidental onely, which naturally does not concern the Place at all; but which may very easily be reformed, without any the least inconvenience, as in due time we shall come to demonstrate.

For first, the City of London is built upon a sweet and most agreeable Eminency of Ground, at the North-side of a goodly and well-conditioned River, towards which it hath an Aspect by a gentle and easie declivity, apt to be improved to all that may render her Palaces, Buildings, and Avenues usefull, gracefull, and most magnificent: The Fumes which exhale from the Waters and lower Grounds lying Southwards, by which means they are perpetually attracted, carried off or dissipated by the Sun, as soon as they are born, and ascend.

Adde to this, that the Soil is universally Gravell, not onely where the City itself is placed, but for several Miles about the Countreys which environ it: That it is plentifully and richly irrigated, and visited with Waters which Christalize her Fountains in every Street, and may be conducted to them in such farther plenty, as Rome herself might not more abound in this liquid ornament, for the pleasure and divertisement, as well as for the use and refreshment of her Inhabitants. I forbear to enlarge upon the rest of the conveniences which this August and

Opulent City enjoies both by Sea and Land to accumulate her Encomiums, and render her the most considerable that the Earth has standing upon her ample bosome; because, it belongs to the Orator and the Poet, and is none of my Institution: But I will infer, that if this goody City justly challenges what is her due, and merits all that can be said to reinforce her Praises, and give her Title; she is to be relieved from that which renders her less healthy, really offends her, and which darkens and eclipses all her other Attributes. And what is all this, but that Hellish and dismall Cloud of SEA-COALE? which is not onely perpetually imminent over her head; For as the Poet,

Conditur in tenebris altum caligine caelum:

but so universally mixed with the otherwise wholesome and exellent Aer, that her Inhabitants breathe nothing but an impure and thick Mist, accompanied with a fuliginous and filthy vapour, which renders them obnoxious to a thousand inconveniences, corrupting the Lungs, and disordering the entire habit of their Bodies; so that Catharrs, Phthisicks, Coughs and Consumptions, rage more in this one City, than in the whole Earth besides.

I shall not here much descant upon the Nature of Smoakes, and other Exhalations from things burnt, which have obtained their several Epithetes, according to the quality of the Matter consumed, because they are generally accounted noxious and unwholesome; and I would not have it thought, that I doe here Fumos vendere, as the world is, or blot paper with insignificant remarks: It was yet haply no inept derivation of that Critick, who took our English, or rather, Saxon appellative, from the Greek word σμύχω corrumpo and exuro, as most agreeable to its destructive effects, especially of what we doe here so much declaim against, since this is certain, that of all the common and familiar materials which emit it, the immoderate use of, and indulgence to Sea-Coale alone in the City of London, exposes it to one of the fowlest Inconveniences and reproaches, than possibly beffall so noble, and otherwise imcomparable City: And that, not from the Culinary fires, which for being weak, and lesse often fed below, is with such ease dispelled and scattered above, as it is hardly at all discernible, but from some few particular Tunnells and Issues, belonging only to Brewers, Diers, Lime-burners, Salt and Sope-boylers, and some other private Trades, One of whose Spiracles alone, does manifestly infect the Aer, more than all the Chimnies of London put together besides. And that this is not the least Hyperbolic let the best of Judges decide it, which I take to be our senses: Whilst these are belching it forth their sooty jaws, the City of London resembles the face rather of Mount AEtna, the Court of Vulcan, Stromboli, or the Suburbs of Hell, than an Assembly of Rational Creatures, and the Imperial seat of our incomparable Monarch. For when in all other places the Aer is most Serene and Pure, it is here Ecclipsed with such a Cloud of Sulphure, as the Sun itself, which gives day to all the World besides, is hardly able to penetrate and impart it here; and the weary Traveller, at many Miles distance, sooner smells, than sees the City to which he repairs. This is that pernicious

Smoake which sullyes all her Glory, superinducing a sooty Crust or Fur upon all that it lights, spoyling the moveables, tarnishing the Plate, Gildings and Furniture, and corroding the very Iron-bars and hardest Stones with these piercing and acrimonious Spirits which accompany its Sulphure; and executing more in one year, than exposed to the pure Aer of the Country it could effect in some hundreds.

> ——— piceaque gravatum
> Faedat nube diem;

It is this horrid Smoake which obscures our Churches, and makes our Palaces look old, which fouls our Clothes, and corrupts the Waters, so as the very Rain, and refreshing Dews which fall in the several Seasons, precipate this impure vapour, which, with its black and tenacious quality, spots and contaminates whatever is exposed to it.

> ——— Calidoque involvitur undique fumo;

It is this which scatters and strews about those black and smutty Atomes upon all things where it comes, insinuating itself into our very secret Cabinets, and most precious Repositories: Finally, it is this which diffuses and spreads a Yellownesse upon our choycest Pictures and Hangings; which does this mischief at home, is Avernus to Foul, and kills our Bees and Flowers abroad, suffering nothing in our Gardens to bud, display themselves, or ripen; so as our Anemonies and many other choycest Flowers, will by no Industry be made to blow in London, or the Precincts of it, unlesse they be raised on a Hot-bed, and governed with extraordinary Artifice to accellerate their springing; imparting a bitter and ungrateful Tast to those few wretched Fruits, which never arriving to their desired maturity, seem, like the Apples of Sodome, to fall even to dust, when they are but touched. Not therefore to be forgotten, is that which was by many observed, that in the year when Newcastle was beseiged and blocked up in our late Wars, so as through the great Dearth and Scarcity of Coales, those famous Works many of them were either left off, or spent but few Coales in comparison to what they now use: Divers Gardens and Orchards, planted even in the very heart of London, (as in particular my Lord Marquesse of Hertford's in the Strand, my Lord Bridgewater's and some others about Barbican) were observed to bear such plentiful and infinite quantities of Fruits, as they never produced the like either before or since, to their great astonishment: but it was by the Owners rightly imputed to the penury of Coales, and the little Smoake, which they took notice to infest them that year: For there is a virtue in the Aer, to penetrate, alter, nourish, yea and to multiply Plants and Fruits, without which no vegetable could possibly thrive: but as the Poet,

> Aret ager: vitio moriens sitit aeris herba:

So as it was not ill said by Paracelsus, that of all things, Aer only could be truly affirmed to have Life, seeing to all things it gave Life. Argument sufficient to demonstrate, how prejudicial it is to the Bodies of Men, for that can never be Aer fit for them to breath in, where not Fruits, nor Flowers do ripen, or come to a seasonable perfection.

I have strangely wondred, and not without some just indignation, when the South-wind has been gently breathing, to have sometimes beheld that stately House and Garden belonging to my Lord of Northumberland, even as far as Whitehall and Westminster, wrapped in a horrid Cloud of this Smoake, issuing from a Brewhouse, or two contiguous to that noble Palace: so as coming up the River, that part of the City has appeared a Sea where no Land was within ken; the same frequently happens from a Lime-kelne on the Banke-side neer the Falcon, which when the Wind blowes Southern, dilates itself all over that Poynt of the Thames, and the opposite part of London, especially about St. Paul, poysoning the Aer with so dark and thick a Fog, as I have been hardly able to pass through it, for the extraordinary stench and halitus it sends forth; and the like is neer Fox-hall at the farther end of Lambeth.

Now to what funest and deadly Accidents the assiduous invasion of this Smoake exposes the numerous Inhabitants, I have already touched, whatsoever some have fondly pretended, not considering that the constant use of the same Aer (be it never so impure) may be consistent with Life and a Valetudinary state; especially, if the Place be native to us, and that we have never lived for any long time out of it; Custome, in this, as in all things else, obtaining another Nature, and all Putrefaction, proceeding from certain Changes, it becomes, as it were, the Form, and Perfection of that which is contained in it: For so (to say nothing of such as by assuefaction have made the rankest poysons their most familiar Diet) we read that Epimenides continued fifty years in a damp Cave, the Eremites dwelt in Dens, and divers live now in the Fens; some are condemned to the Mines, and others, that are perpetually conversant about the Forges, Fornaces of Iron and other Smoaky Works, are little concerned with these troublesome accidents: But as it is not (I perswade myself) out of choyce, that these Men affect them; so nor will any man, I think, command and celebrate their manner of Living. A Tabid Body might possibly trail out a miserable Life of seven or eight years by a Sea-coale Fire, as 'tis reported the Wife of a certain famous Physician did of late, by the Prescription of her Husband; but it is to be considered also, how much longer, and happier she might have survived in a better and more noble Aer; and that old Par, who lived in health to an Hundred and fifty years of Age, was not so much concerned with the change of Diet (as some have affirmed) as with that of the Aer, which plainly withered him, and spoyled his Digestion in a short time after his arrival at London.

There is, I confesse, a certain Idiosyncrasia in the Composition of some persons, which may fit and dispose them to thrive better in some Aers, than in other: But it is manifest, that those who repair to London, no sooner enter into it, but they find a universal alteration in their Bodies, which are either dryed up or enflamed, the humours being exasperated and made apt to putrifie, their sensories and perspiration so exceedingly stopped, with the losse of Appetite, and kind of general stupefaction, succeeded with such Catharrs and Distillations, as do never, or very rarely quit them, without some further Symptomes of dangerous Inconveniency so long as they abide in the place; which yet are immediately restored to their former habit, so soon as they are retired to their Homes and enjoy the fresh Aer again. And here I may not omit to mention what a most Learned Physician, and one of the College, assured me, as I remember of a Friend of his, who had so strange an Antipathy to the Aer of London; that though he were a Merchant, and had frequent businesse in the City, was yet constrained to make his Dwelling some miles without it; and when he came to the Exchange, within an hour or two, grew so extremely indisposed, that (as if out of his proper Element) he was forced to take horse (which used therefore constantly to attend him at the Entrance) and ride as for his Life, till he came into the Fields, and was returning home again, which is an Instance so extraordinary, as not, it may be, to be paralell'd in any place of Europe, save the Grotto del Cane, nere Naples, the Os Plutonium of Silvius, or some such subterranean habitation. For diseases proceed not from so long a Series of causes, as we are apt to conceive; but, most times from those obvious, and despicable mischiefs, which yet we take lesse notice of, because they are familiar: But how frequently do we hear men say (speaking of some deceased Neighbour of Friend) He went up to London, and took a great Cold, &c., which he could never afterwards claw off again.

I report myself to all those who (during these sad confusions) have been compelled to breath the Aer of other Countries for some years, if they do not now perceive a manifest alteration in their Appetite, and clearance of their Spirits; especially such as have liv'd long in France, and the City of Paris; where, to take off that unjust reproach, the Plague as seldome domineers, as in any part of Europe, which I more impute to the Serenity and Purity of the Aer about it, than to any other qualities which are frequently assign'd for the cause of it by divers writers. But if it be objected that the purest Aers are soonest infected; it is answered, that they are also the soonest freed again; and that none would therefore choose to live in a corrupt Aer, because of this Article: London, 'tis confess'd, is not the only City most obnoxious to the Pestilence; but it is yet never clear of this Smoake which is a Plague so many other ways, and indeed intolerable; because it kills not at once, but always, since still to languish, is worse than even Death itself. For is there under Heaven such Coughing and Snuffing to be heard, as in London Churches and Assemblies of People, where the Barking and the Spitting is uncessant and most importunate. What shall I say?

Hinc hominum pecudumque Lues

And what may be the cause of these troublesome effects, but the inspiration of this infernal vapour, accompanying the Aer, which first heats and sollicits the Aspera Arteria, through one of whose Conduits, partly Cartilaginous, and partly

Membranous, it enters by several branches into the very Parenchyma, and substance of the Lungs, violating, in this passage, the Larynx and Epiglottis, together with those multiform and curious Muscles, the immediate and proper Instruments of the Voyce, which becoming rough and drye, can neither be contracted, or dilated for the due modulation of the Voyce; so as by some of my Friends (studious in Musick, and whereof one is a Doctor of Physick) it has been constantly observ'd, that coming out of the Country into London, they lost Three whole Notes in the compass of their Voice, which they never recover'd again till their retreat; Adeo enim Animantes (to use the Orators words) aspiratione Aeris sustinentur, ipseque Aer nobiscum videt, nobiscum audit, nobiscum sonat: In summe, we perform nothing without it.

Whether the Head and the Brain (as some have imagined) take in the ambient Aer, nay the very Arteries) through the skin universally over the whole body, is greatly controverted; But if so, of what consequence the goodnesse and purity of the Aer is, will to every one appear: Sure we are, how much the Respiration is perturb'd, and concern'd when the Lungs are prepossessed with the grosse and dense vapours, brought along in the Aer; which on the other side being pure and fitly qualified, and so conducted to them, is there commixed with the circulating blood, insinuating itself into the left ventricle of the heart by the Arteria Venosa, to rarifie and subtilize that precious vehicle of the Spirits and vital flame: The Vena Arteriosa, and Arteria Venosa, disposing themselves into many branches through the Pulmonique lobes, for its Convoy of the Aer (as we sayd) being first brought into them out of the Bronchia (together with the returning blood) to the very Heart itself; so as we are not at all to wonder, at the suddain and prodigious Effects of a poysonous or lesse wholesome Aer, when it comes to invade such noble Parts, Vessells, Spirits and Humours, as it visits and attaques, through those subtitle and curious passages. But this is not all.

What if there appear to be an Arsenical vapour, as well as Sulphur, breathing sometimes from this intemperate use of Sea-Coale, in great Cities? That there is, what does plainly stupifie, is evident to those who sit long by it; and that which fortun'd to the Dutchman who winter'd in Nova Zembla, was by all Physicians attributed to such a deleterious quality in the like fuel, as well as to the Inspissation of the Aer, which they thought only to have attemper'd as is by most esteem'd to be the reason of the same dangerous halitus of Char-Coale, not fully enkendl'd. But to come nearer yet.

New Castle Coale, as an expert Physician affirms, causeth Consumptions, Phthisicks, and the Indisposition of the Lungs, not only by the suffocating aboundance of Smoake; but also by its Virulency; For all subterrany Fuell hath a kind of Virulent or Arsencial vapour rising from it, which as it speedily destroys those who dig it in the Mines; so does it by little and little, those who use it here above them: Therefore those Diseases (saith this Doctor) most affict about London, where the very Iron is sooner consum'd by the Smoake thereof, than where this Fire is not used.

And, if indeed there be such a Venemous quality latent, and sometimes breathing from this Fuell, we are lesse to trouble our selves for the finding out of the Cause of those Pestilential and Epidemical Sicknesses (Epidemiorum Causa enim in Aere, says Galen) which at divers periods, have so terribly infested and wasted us: or, that it should be so susceptible of infection, all manner of Diseases having so universal a vehicle as is that of the Smoake, which perpetually invests this City: But this is also noted by the Learned Sir Kenelme Digby, in confirmation of the doctrine of Atomical Effluvias and Emanations, wafted, mixed and communicated, by the Aer, where he well observes, that from the Materials of our London Fires, there results a great quantity of volatile Salts, which being very sharp and dissipated by the Smoake doth infect the Aer, and so incorporate with it, that though the very Bodies of those corrosive particles escape our perception, yet we soon find their effects, by the destruction which they induce upon all things that they do but touch; spoyling, and destroying their beautiful colours, with their fuliginous qualities: Yea, though a Chamber be never so closely locked up, Men find at their return, all things that are in it, even covered with a black thin Soot, and all the rest of the Furniture as full of it, as if it were in the house of some Miller, or a Baker's Shop, where the Flower gets into their Cupboards, and Boxes, though never so close and accurately shut.

This Coale, says Sir K. flies abroad, fowling the Clothes that are expos'd a drying upon the Hedges; and in the Spring-time (as but now we mention'd) besoots all the Leaves, so as there is nothing free from its universal contamination and it is for this, that the Bleachers about Harlem, prohibit by an express Law (as I am told) the use of these Coales, for some Miles about that Town; and how curious the Diers and Weavers of Dammask, and other precious Silks are at Florence, of the least ingresse of any Smoaky vapour, whilst their Loomes are at work, I shall shew upon some other occasion: But in the mean time being thus incorporated with the very Aer, which ministers to the necessary respiration of our Lungs, the Inhabitants of London, and such as frequent it, find it in all their Expectorations; the Spittle, and other excrements which proceed from them, being for the most part of a blackish and fuliginous Colour: Besides this acrimonious Soot produces another sad effect, by rendring the people obnoxious to Inflammations, and comes (in time) to exulcerate the Lungs, which is a mischief so incurable, that it carries away multitudes by languishing and deep Consumptions, as the Bills of Mortality do Weekly inform us. And these are those Endemii Morbi, vernaculous and proper to London. So corrosive is this Smoake about the City, that if one would hang up Gammons of Bacon, Beefe, or other Flesh to fume, and prepare it in the Chimnies, as the good House-wifes do in the Country, where they make use of sweeter Fuell, it will so Mummifie, dry up, wast and burn it, that it suddainly crumbles away, consumes and comes to nothing.

The Consequences then of all this is, that (as was said) almost one half of them who perish in London, dye of Phthisical and Pulmonic distempers; That the Inhabitants are never free from Coughs and importunate Rheumatisms spitting of Impostumated and corrupt matter: for remedy whereof, there is none so infallible, as that, in time, the Patient change his Aer, and remove into the Country: Such as repair to Paris (where it is excellent) and other like Places, perfectly recovering of their health; which is a demonstration sufficient to confirm what we have asserted, concerning the perniciousnesse of that about this City, produc'd only, from this exital and intolerable Accident.

But I hear it now objected by some, that in publishing this Invective against the Smoake of London, I hazard the engaging of a whole Faculty against me, and particularly, that the Colledge of Physicians esteem it rather a Preservation against Infections, than otherwise any cause of the sad effects which I have enumerated. But, as I have upon several encounters, found the most able, and Learned amongst them, to renounce this opinion, and heartily wish for a universal purgation of the Aer by the expedients I propose; so, I cannot believe that any of that Learned Society, should think themselves so far concerned, as to be offended with me for that, which (as well for their sakes, as the rest who derive benefit from it) I wish were at farther distance; since it is certain, that so many of their Patients are driven away from the City, upon the least indisposition which attaques them, on this sole consideration; as esteeming it lesse dangerous to put themselves into the hands of some Country Doctor or Emperic, then to abide the Aer of London, with all its other advantages. For the rest, that pretend to that honourable Profession; if any shall find themselves agreev'd and think good to contend, I shall easily allow him as much Smoake as he desires, and much good may it do him. But, it is to be suspected and the answer is made (by as many as have ever suggested the Objection to me) That there be some whom I must expect to plead for that, which makes so much work for the Chimny-Sweeper; Since I am secure of the Learned and Ingenuous, and whose Fortunes are not built on Smoake, or raised by a universal Calamity; such as I esteem to be the Nuisances, I have here reproved: I do not hence infer, that I shall be any way impatient of a just and civil Reply, which I shall rather esteem for an honour done me, because I know, that a witty and a learned man us able to discourse upon any Subject whatsoever; some of them having with praise, written even of the praise of Diseases themsleves, for so Favorinus of old, and Menapius since commended a Quartan Ague, Pirckhemierus the Gout, Gutherius celebrated Blindnesse, Heinsius the Louse, and to come nearer our Theam, Majoragius the nasty Dirt; Not I suppose that they affected these pleasant things, but as A. Gellius has it, exercendi gratia, and to shew their Wits; for as the Poet,

> Sunt etiam Musis sua ludicra, mista Camenis
> Otia sunt:

But to proceed, I do farther affirm, that it is not the Dust and Ordure which is daily cast out of their Houses, much lesse what is brought in by the

Feet of Men and Horses; or the want of more frequent and better conveyances, which renders the Streets of London dirty even to a Proverb: but chiefly this continual Smoake, which ascending in the day-time, is, by the descending Dew, and Cold, precipitated again at night: And this is manifest, if a piece of clean Linnen be spread all Night in any Court or Garden, the least infested as to appearance: but especially if it happens to rain, which carries it down in greater proportion, not only upon the earth, but upon the Water also, where it leaves a thin web, or pellicule of dust, dancing upon the Surface of it; as those who go to bathe in the Thames (though at some Miles distant from the City) do easily discern and bring home upon their Bodies: How it sticks on the Hands, Faces and Linnen of our fair Ladies, and nicer Dames, who reside constantly in London (especially during Winter) the prodigious wast of Almond-power for the One, Soap and wearing out of the Other, do sufficiently manifest.

Let it be considered what a Fuliginous crust is yearly contracted, and adheres to the Sides of our ordinary Chimnies where this grosse Fuell is used; and then imagine, if there were a solid Tentorium, or Canopy over London, what a masse of Soote would then stick to it, which now (as was said) comes down every Night in the Streets, on our Houses, and Waters, and is taken into our Bodies.

And may this much suffice concerning the Causes and Effects of this Evill, and to discover to all the World, how pernicious this Smoake is to our Inhabitants of London, to decrie it, and to introduce some happy Expedient, whereby they may for the Future, hope to be freed from so intolerable an inconvenience, if what I shall be able to produce and offer next, may in some measure contribute to it.

An Insanitary Era*

J. C. Wylie

When Elizabeth I came to the throne the population of the country was around four millions and in the course of her reign London doubled in size to reach two hundred thousand. By then it had outstripped Venice and stood unrivalled as the largest city in the world. There was nothing like it in all England, where an average town of the time could muster no more than five thousand citizens. But in larger cities like York, Norwich and Bristol all of which had already expanded to twenty thousand, as well as in London, many of the problems which are nowadays so closely associated with the rapid growth of towns were beginning to reveal themselves. Great quantities of food had to be brought in to feed people who were engaged on work which was far removed from agriculture; environmental services were also needed to keep the people in health.

The Romans taught us the value of personal and environmental hygiene but what they taught was soon forgotten. The people the Romans left behind in Britain thought nothing of the sanitary systems they had been given and thereby proved themselves to be no more civilized than the Huns and the Goths who threatened Rome. The Roman villa in Britain had been a farmstead fitted with baths and central heating, but when Henry III came to the throne he found his privy built into the thickness of his castle's wall with a discharge into the moat. He found it too smelly and inconvenient for use and was glad to pay the substantial sum of one hundred pounds for a new one. Things were no better at the time of Elizabeth; during her reign, Sir William Harington, a manufacturer of garden fountains, produced a design for a water-closet, but was considered eccentric. Even our immediate grandfathers were not too fastidious and an appreciation of cleanliness by modern standards is not as old as the internal combustion engine. Queen Victoria was in fact the first reigning monarch of Britain to have a properly equipped bathroom installed in one of her castles and today there are more television sets than baths in Chicago.

It was fortunate that the medieval town, though surrounded by a wall, was still part of the country. The only glimpse some Londoners get nowadays of the country is when they move once a year into Kent to help with the hop-picking, but in those far-off days the whole population of towns and villages moved into

*Reprinted by permission of Faber and Faber, Ltd. from <u>The Wastes of Civilization</u>.

the country every year to help with the harvest on which they depended for food. Fowl, game and fish could be had by all, for the citizens of London had the right to hunt and fish over wide areas of surrounding country. All this gave Londoners a sound basis for health. They greatly needed it, for the only scavengers of medieval towns were pigs, dogs and kites, that almost extinct bird which is the nearest our climate will allow to the vulture. The dogs have survived, and pigs were used for scavenging in Manchester as late as the middle of the nineteenth century, but flocks of animals wandering through the streets of towns create their own nuisance.

In Norman times, swine were allowed to foul the partially cobbled streets as they were driven to pastures beyond the town walls, but it was not long before 'the keepers of pigs and geese' were being ordered to keep their stock indoors, and butchers who often occupied a whole street called 'The Shambles' were being made to remove the blood and offal of their trade. The streams on which many towns were built were used to carry away the litter and debris. Many of these have since been covered over to serve as sewers and today only a few remain to flow alongside shopping streets where Woolworth buildings and multiple stores display wireless and television sets, and bakers' shops sell wrapped bread.

The first Sanitary Act in British History was passed by Parliament sitting in Cambridge in 1388. It provided that the townspeople themselves were 'to remove from the street and lanes of towns all swine and all dirt, filth and branches of trees and to cause the streets and lanes to be kept clean for the future'. But towns cannot be kept clean by statute alone for it was impractical for every citizen to wait till after nine o'clock at night before disposing of his 'dung or goung' as he was directed. Instead, this matter was deposited in the streets under cover of darkness. So it went on through the centuries, secretly at first and then more blatantly as the task, with the growth of towns, became increasingly difficult.

In 1512 a Mayor of Nottingham was summoned for selling rotten herring and for starting a muck-hill; in 1690 a Mayor of Portsmouth was fined for throwing garbage into the streets of his own town; there were riots in Paris during the eighteenth century because the mass of filth littering the streets was more than an indignant people could endure; in Edinburgh, the warning shout of 'gardeloo' with its responding cri de coeur 'haud yer haun' continued to be heard till the close of the eighteenth century, whenever a tenement window was opened and all manner of liquid and solid filth was thrown into the streets; in 1805 a Lord Mayor of Dublin issued a proclamation, instructing that, in order to protect the people from the diseases incidental to filth, he would indemnify them for removing it and it is reported that 'one gentleman in consequence of the notification, cleansed seven perches (about 40 yards) of street about his house and in doing so removed one hundred and fifty loads of gutter'.

Even one hundred years ago there was still argument as to who should be responsible for the cleaning of streets and in September 1856 The Economist commented: 'Amongst the matters recently referred to by the Registrar General as affecting the health of the metropolis are the cleansing of the streets, now covered with horse dung. The subject has two distinct aspects, the material and the moral, or what should be done with the refuse matter we must get rid of or perish, and who should do it.' Fifty years later the hills of ordure in the town of Paisley were being defended by citizens armed with pistols, for some of them would rather endure the smell and endanger their health, than lose the few shillings that farmers were prepared to pay them for their 'fulzie'.

However, the emergence from the Dark Ages did bring a mild sanitary renaissance. Private baths with adjoining dressingrooms reappeared during the thirteenth century and for a time the bath-house came to serve again as it had in Roman times, but, as towns grew in size, family life deteriorated and the bath-house degenerated to a place for gossip and loose living. Nevertheless, in its remedial measures for health the medieval town was far in advance of the industrial towns of the nineteenth century which succeeded it. The Holy Orders founded hospitals which gave the sick a service which, for the time, was as adequate as anything provided today by the Welfare State.

In London privies were still being built over the Fleet but the Wallbrook was bricked over to give the city its first public sewer since Roman times. In 1307 the first stone bridge over the Thames was completed, carrying houses several stores high, as well as traffic. From the houses, rubbish was thrown into the Thames, which was already being described as a 'great street paved with water and filled with shipping'. The same people who fouled the river lowered buckets from their windows to collect drinking water. Edward I appointed surveyors to stop litter being thrown on the roadway and into the river, and the merchants who entered the city by the bridge in addition to paying pontage for the right to use the bridge, paid 'stallage' as a rent for their stance at the market, which also covered the cost of removing their rubbish.

John of Gaunt, a patron of the arts, made Chaucer responsible for seeing that the ditches and drains between Greenwich and Woolwich were kept in good repair. The appointment was a sinecure but it gave him an income and we can be grateful that it gave the poet much spare time to devote to his writing even if it could add little to the comfort of his fellow citizens. The collection of refuse was also being organized and men could be conscripted for the menial task of scavenging. Each ward in London had its 'rayker' and refuse could be put out for collection on the days the cart came round. But only twelve rubbish carts served the whole city and most of the rubbish was thrown into the streets to be collected and transferred to laystalls in ground outside the city walls or on the banks of the Thames, whence dung-boats carried it along the river to be disposed of by methods which remain a mystery.

The streets had no footpaths and were badly paved, sloping from the crown to 'kennels' into which the filth ran. The weakest pedestrian was 'pushed to the wall' where he splashed through the mud. Still when Elizabeth came to the throne, London was no more than a group of villages, interspersed with woods and surrounded by forest and heath. Hyde Park was a country district and birds still sang in Charing Cross. There were not more than four million people in the whole country and simple sanitary arrangements served tolerably well. The people enjoyed reasonably good health, and the sporadic outbreaks of plague were not worse than the visitations of cholera, typhoid and dysentery of the nineteenth century or influenza and poliomyelitis of the present century.

Nothing bigger than a hundred-ton ship came up the Thames, and although Members of Parliament could fish for salmon from the terrace of the House of Commons, the water of the river was barely fit to drink. A patent for a leaden conduit for conveying water from Tyborne Brook to the City of London was granted in 1236, but at the beginning of the seventeenth century the citizens of London were almost entirely dependent for drinking water on what was carried to them from the Thames by the Honourable Company of Water Tankard Bearers, who had the right to sell the water under charter. When the New River Company, also under charter, first brought clean water into the city in 1610 and supplied it to houses in pipes made of elm wood, a serious industrial dispute arose, which only the passing of time resolved. The activity of the New River Company deprived the Tankard Bearers of their living, but their living could not be protected in the face of the convenience of the new service and the increasing pollution of the Thames.

Sir Francis Drake took a hand in the development of public water supplies, but his motives were, like many of his exploits on the high seas, not above suspicion. He used his influence and persuasive powers in Parliament where he was a Member, to get authority to carry water from the River Meavy to Plymouth. Having got the powers he sought, which included making use of the water for industrial purposes, he operated the only mill he owned off the newly installed supply and then expanded his business. He erected additional mills until he was using all the available water and then left the people of Plymouth no better off than they were before he interested himself in their welfare.

By this time a more regular use was being made of town ordure for manuring. The practice of manuring was known in England in the fourteenth century and by the end of the sixteenth century it had been brought to a fine and imaginative art. Chalk was mined; sand from the coast was brought in to lighten heavy clay; sea-weed was spread and ploughed in raw or after being composted in heaps; every form of dung, animal and human, along with industrial wastes like soot, rope waste, rags, hair, malt waste, dust and bark were used whenever they could be had. Manure was described as 'almost anything that hath liquidnesse, foulnesse, saltnesse or good moisture in it' and the effete matter of a population was found to be just the thing for manuring land from which its food was grown. A Lord Provost of the City of Edinburgh of the time carried his civic duty to the point where he contracted to take all the city's ordure and used it to manure the fields of his Prestonfield Estate. These lands soon acquired and still hold a high reputation for fertility. It was common practice throughout the country for farmers to collect city ordure as a return load when they carted food to market, or fodder to city stables and dairies, and it is recorded that the malt barges that plied down the Lea from Ware came back loaded with dung for the light chalk uplands of Hertfordshire.

But the city could not be kept clean by the unorganized efforts of farmers alone and while the conditions in the streets were like those of a stockyard, the rush floored halls of Tudor England, with the trodden earth and the dogs, where stale rushes and bones mingled with decaying food and debris, were no cleaner than the bottom of a palaeolithic cave. This debris was dug out less often than a modern deep-litter poultry house. When there was a thickness of two or three feet it was removed to caves where it was mixed with urine, blood and wood ash — just the thing for manuring fields and growing good crops; instead, after being allowed to mature for two years, this matter was used to manufacture gunpowder.

The discovery of gunpowder has been credited to many. The Germans stake a claim for it but Friar Roger Bacon, a medieval alchemist, born at Ilchester in Somerset in 1214, who devoted himself to religion and the search for the elixir, is a worthy even if he is a less likely candidate. In his book De Secretia, Friar Bacon attacks magic on the grounds that science and the arts could, even at that time, reveal greater wonders than all the alleged revelations of the Black Art of Magic. The first eight chapters of his book deal with things which he believed art could produce and magic could not. In the next three chapters he conceals, in a confusion of words, his greatest secret, which was a recipe for the manufacture of gunpowder. In this he had the same concern for the use that might be made of his discovery as modern scientists have for the hydrogen and atom bombs, for he says 'the mob scoffs at philosophers and despises scientific truth. If by chance they lay hold upon some great principle, they are sure to misinterpret and misapply it so that what would be gain to everyone causes loss to all. It is madness to commit a secret to writing, unless it be so done as to be unintelligible to the ignorant, and only just intelligible to the best educated.'

Friar Bacon's recipe for gunpowder was 41.2 per cent saltpetre (potassium nitrate), 29.4 per cent sulphur, and 29.4 per cent charcoal. Mixtures of this type sent the first smoke mushrooms in history jetting from the mouths of very small cannon at the battle of Crecy in 1346. By the time Drake sailed to meet the Armada the explosive mixture had been levelled out to 75 per cent saltpetre, 10 per cent sulphur and 15 per cent charcoal and thereafter remained unchanged. Nelson's three-deckers had bigger and better guns at Trafalgar, but roughly the same gunpowder. The last two ingredients are easily procured. Charcoal can be made from any wood though elder and hazel are best, and sulphur, a volcanic product which can be obtained in small quantities in almost any country, could be mined almost pure in Italy and Sicily. Any fertile soil contains nitrogen, fixed from the air by bacteria, both free living (azotobacter) and those in the roots of pea-tribe plants (legumes); soils which contain liberal quantities of dung and urine will contain more than normal amounts.

All London was built over fertile soil and the customs of an insanitary age made the soil around a rich mine of saltpetre. Pigeons were kept, mostly at ground level, and were more popular then than hens are today, and they provided an important augmentation of the raw materials required for the formation of nitrate crystals. Drake could not have felt other than grateful to these pigeons for the supplies of gunpowder they gave him, but now their countless descendants add to cleansing problems in cities, and show scant respect for his illustrious counterpart of a later century by fouling the column erected to his memory in Trafalgar Square with the very substance that earned them his gratitude.

In accordance with the custom of an age when the Civil Service was in its infancy, the right to 'digge for saltpetre' was sold to monopolists. These men, of whom George Evelyn, uncle of the diarist, held the Royal Licence from 1565 to 1624, employed the diggers and ran the powder-mills, the first of which was built in London in 1461. So rich in nitrogen were the floors of Tudor houses that up to 1634 it was illegal for floors to be tiled or paved and so vital to the safety of the country was the nitrate-rich litter that 'saltpetremen' could go in and dig it without obtaining permission from the householder.

The Guildhall records of the time contain many expressions of indignation at this invasion of privacy, which was to be repeated three hundred and fifty years later when the last war proved so demanding in iron that essential things like protective railings were confiscated to be used in the manufacture of weapons of war. In 1601 there was a heated debate in Parliament on the subject of the activities of the saltpetremen. Said one member, 'They digge in dove cotes when the doves are nesting; cast up malting floors when the malt be green, in bedchambers, in sick rooms, not even sparing women in childbed, yea even in God's house, the Church.' But there was to be no quick removal of the grievances for it was explained that the saltpetremen could not be restrained because 'the kingdom is not so well placed for powder as it should be.'

Three years later, Guy Fawkes would get the 100-lb. barrels of gunpowder he placed in the vaults of the House of Commons for ninepence a pound from any one of a dozen powder-mills operating at the time. This powder was still 75 per cent saltpetre won from the floors of Londoners. The first trial consignment from the natural deposits of India did not arrive until 1610, and it was not until 1656 that the East India Company managed to build up and stockpile enough to allow an Act to be passed forbidding any man to enter another's dwelling to dig for saltpetre without prior permission. So as the Englishman's house at last became his castle again one obstacle standing in the way of sanitary reform was removed.

Despite the activities of the farmers and saltpetremen, Pepys found conditions in London worse than anything Chaucer could describe. He complained that the town ditch, which had proved to be too nauseous for the none too fastidious senses of the Elizabethan citizen, could not have been worse than the Fleet River, which he saw as an uncovered sewer of outrageous filthiness. There were still no side walks and the egg-shaped cobbles which paved the streets caused excessive noise, and the intolerable din, he said, brought evil consequences to trade, health and government.

The Great Fire of London of 1666 is a milestone in British sanitary history for it marks the last great outbreak of plague which the country was to see. Yet the Great Plague which struck London a year earlier with such violence that the living were too few to bury the dead, was merely the last of a series of similar outbreaks which had occurred regularly throughout Europe during the preceding four hundred years, and it was probably not the worst. An outbreak in the fourteenth century had caused thirty thousand deaths in England alone, and throughout Europe, at a time of intellectual triumph, brought a break in the continuity of advancement and a lowering of the vitality of the people of a continent equal to that following an exhausting war. In England at the time of the fire there had been immunity from the plague for thirty years and there had also been the Renaissance. It was a time of advance and luxury, and this new outbreak struck the imagination of the people and brought a realization of the dangers of an insanitary environment.

Fire is a good sterilizing agent and the frequent fires which occurred in flimsy medieval houses did much to constrain the tragic effects of the earlier outbreaks. But unless the conditions under which disease spreads are remedied, fire can do no more than provide a temporary check. The Great Fire gave Wren the opportunity to express his genius as architect and town planner, but in the absence of improved santiary arrangements the credit for bringing an end to outbreaks of plague must go to the brown rat. During centuries of trading the brown rat had been brought into this country in countless numbers as stowaways in Britain's expanding fleet of merchantmen. It found conditions which suited it, just as the rabbit did when it first set foot on Australian soil, and being just as prolific it soon overran the countryside.

On farms, cats were bred to keep the rats at bay, but soon it was estimated that the cats were outnumbered by twenty to one. Special diets of milk failed to give the cats sufficient courage to carry on the unequal struggle, while the rats continued to thrive in the good living quarters they discovered in the houses and on the plentiful supply of food they found amongst the filth which lay about in an insanitary age. They even nested in the monstrous wigs of the nobility, and with everything in their favour they not only defied the cats but at the same time extirpated their cousin the medieval black rat, just as the foreign grey squirrel is doing to the native red today. The greatest improvement up till then effected in man's urban environment was thus achieved by a verminous creature with the single virtue that it did not carry the plague flea as the native black rat did. By killing off the black rat, the brown rat put an end to the plague, though it also retarded sanitary progress.

London could not be rebuilt in a day, and as the years passed without a recurrence of the plague the initial enthusiasm for costly and thorough schemes of reconstruction waned. The Commissioners of Sewers who were appointed to give London an all-embracing sanitary authority did little more than paper work and continued in being for more than two centuries without accomplishing much. Under the Rebuilding Acts of 1667 houses of brick were built and the substitution of carpets and wood panelling in place of rush flooring and cloth hangings, effected some sanitary improvement by depriving the carrier fleas and vermin of their harbourage. Despite this, the houses in London at the time of Anne were, in the opinion of Defoe, no better than they were before the fire. With a population approaching three-quarters of a million, London was then a city of contrasting wealth and poverty. Alongside the London which was the centre of the arts, law, government and fashion, there was the squalor of the poorer districts where the people were without police, education or medical attention. The city was not yet left at week-ends to caretakers and cats, but it could no longer be said that those living in London had one foot in the country.

Combined with this poverty there was still greater wealth. Inevitably there was more to throw away. Bones, feathers from game, fish waste and the outside leaves of vegetables made up the decayable mass. Broken china and broken bottles would be more frequently discarded by the upper classes, the poorer classes being still served by wooden platters and the leather bottle, which now only survive on inn signs. These did not break and metal spoons were too precious not to be treasured. Lost coins, however, would be found, even forgotten hoards and sometimes mystifying things. Paper and cardboard were still scarce and costly; nothing was wrapped or tinned. Newspapers had not yet replaced the news-letters. The Times did not appear until 1785 and it was still later before its circulation reached five thousand. Even the grounds from Lloyd's Coffee House or the leaves from Dr. Johnson's dish of tea would not be in any quantity, and time was still needed for the peel of potatoes to appear in the quantity it did when the potato became the staple diet of industrial Britain, and Cobbett, with good reason, was moved to call it 'this root of misery.'

But more coal was being burned to increase the volume of refuse which householders discarded. Outcrop coal was used for domestic fires even before the Roman occupation. King Edward I had a monopoly on the sale of 'sea-cole' brought into London in 'keel boats' from Northumberland. In 1306, Parliament petitioned him to stop the burning of this coal in London because they considered the fumes a danger to health. Pepys complained about the 'smoak and aire' of London and Sir John Evelyn was even more vociferous. He was indignant that London 'should wrap her stately head in clouds of smoke and sulphur, so full of stink and darkness' which he said emanated from 'some few particular tunnels and issues, belonging only to brewers, dyers, lime-burners, salt- and soap-boilers and some other private trades.'

Today, after centuries of discomfort and acute suffering lasting more than one hundred years, old people are being advised to stay indoors during the smog-ridden months of winter while a few health authorities cautiously introduce remedial measures. In 1952, four thousand people died in one London pea-souper and the best beef cattle in the country, brought in for the Smithfield Show, were killed off in their stalls as effectively as if they had been attacked with poison gas. Thirty-seven people in a hundred in London are needlessly afflicted with respiratory disease and forty of these die of such disease every week. In Birmingham, Manchester, Liverpool, Glasgow and in every industrial town of any size in the country, people suffer in the same way. The cost of this plague is estimated at three hundred million pounds a year but the cost of having things put right is no more than the extra trouble involved in burning our dwindling stocks of coal more efficiently than we do at present.

By the time of the Restoration about a million tons of coal a year were being mined and the first great gulf between the classes in British social history was appearing, for the medieval peasants and artisans were not segregated from their neighbours as were the increasing number of miners in the middle of the seventeenth century. It is true that without decent roads and an organized transport system the coal won could not be distributed properly; throughout the country people had to rely on what wood they could glean to keep themselves warm and dried dung was still being burned in many parts of the country. Nevertheless during the eighteenth century coal production shot up from three million tons to over ten million tons a year; not much against the two hundred and fifty million tons output in 1900 but still enough to increase the bulk of rubbish and create a new sanitary problem.

Coal ash has a binding quality and as layer upon layer of refuse with an increasing ash content was thrown on to the streets, each layer was consolidated by passing traffic and the level of the streets was gradually raised. Excavations in some streets in Edinburgh have revealed that modern heavy traffic is carried on a thickness of consolidated refuse which sometimes reaches a depth of fifteen feet accumulated during these insanitary years, and basements, a floor's depth below road level, which are a common feature of ancient cities such as Edinburgh, are thus explained.

As the people of Britain entered upon an era of unprecedented industrial development, an Empire and rapidly expanding trade abroad gave access to almost unlimited food; they were not in danger of eating themselves out of existence as the Mayas had done. But this increasing wealth brought more rubbish, and with arrangements for disposing of it no more adequate than those of medieval town, the fop of the eighteenth century, with his red-heeled shoes, had to tread even more delicately than his counterpart of earlier ages, when he contemplated the wretch in the pillory being pelted with dead cats and turnips.

The very wealth which was Britain's demanded organized sanitary services to rescue the people from the ignominious fate of being buried alive in their own refuse and to protect them from the ominous threat of disease more lethal than the plague.

Foul Seas and
Polluted Streams *

J. C. Wylie

With more than four-fifths of the population in this island living within easy reach of the coast, it is comparatively easy to get rid of sewage by discharging it into the sea. But the assumption that the sea and tidal waters provide limitless dilution and consequently a cheap and effective method of sewage disposal is illusory. Fifty years ago coastal towns were not the popular holiday resorts they are today. Then, the great bulk of the population could not afford to go on holiday and the pollution of beaches did not matter much. Now everyone goes on holiday and half the population of the country seems to spend the summer months at the seaside, to catch what sun is going and to suffer pollution coming at them from all directions, and it is being seriously said that polluted bathing waters are causing poliomyelitis just as contaminated water supplies caused cholera one hundred years ago.

Out at sea, the world's shipping dumps used oil into the sea by the ton. Sometimes whole tanker loads have to be discharged to form floating islands of oil covering hundreds of square miles of ocean. Flame-throwers have been used to burn the oil and with it the carcasses and bodies of thousands of birds, dead, or suffering an agonizing death. But once dumped, it is impossible to destroy the floating oil and not all maritime countries are prepared to legislate against the discharging of waste oil off territorial waters. Without international agreements, no single country can protect its own shores against this menace and globules of oil will continue to be washed up on every beach throughout the world, to be trodden on by people looking for recreation. The oil is not lethal, but it ruins clothes, and on the more polluted beaches, a new trade in the sale of solvents has sprung up, which may yet pay better than the ice cream or paper hat trades.

Solvents are wonderfully effective in removing the filthy oil picked up on feet and legs, and the public which buys them is conscious of the disagreeable conditions which make them necessary. But the harm done when sewage is discharged offshore is neither so easily put right nor is it so apparent. Special

*Reprinted by permission of Faber and Faber Ltd. from The Wastes of Civilization.

means have to be devised to trace the mischief being done. Sewage is always at a higher temperature than sea-water, so that no matter how it is discharged it comes to the surface, bringing with it faecal matter and floating debris. On coming to the surface these are at the mercy of wind and currents and tend to be carried back to the coast by incoming tides.

Patches of oil and brightly coloured floats have been used to establish the line of drift of floating matter, but these are not altogether satisfactory as they are affected by wind to a greater extent than the large volume of finely suspended organic matter contained in sewage, which usually lies just below the surface of the water. A type of bacteria called Serratia indica has been used to obtain more accurate readings. These bacteria will live in sewage for forty-eight hours and in sea-water for five days, and they possess the useful attribute of developing a characteristic and easily visible red colour. Still more delicate tests have been carried out using radio-active phosphorus, which can be detected with great sensitivity by means of Geiger counters. By this latter method it has been established that much of the sewage sludge which is daily discharged from barges off the mouth of the Thames surges backwards and forwards with the tides for days, before being finally deposited on the banks of the river, which are now more nauseous, and that along a greater mileage, than they were when they provided a living for the wretched mud-rakers of Victorian times.

Health authorities who discharge crude sewage into the sea may preen themselves in the light of the cheap disposal service they operate, but they should know that, in leaving bathing under such conditions to the discretion of the individual they are endangering the health of the very people whose health they are supposed to protect, and there is no good reason why a town on the seaboard should spend less on treating sewage than any other town. In some States in America the purity of sea-water is tested and if it is shown to contain more than 50 Coli Bacilli per 100 c.c., bathing is prohibited. The introduction of similar legislation in this country would seriously embarrass most health authorities of popular holiday resorts, and boardinghouse keepers and tradesmen would then gladly pay the few extra pence on their rates required for the proper treatment of their sewage before it is discharged in the sea.

A protracted series of float tests were recently carried out around the coast of the island of Jersey which established that, no matter where sewage was discharged off the island, it was invariably carried round the coast and deposited on the largest and most popular bathing beach. The bathing pool, which was automatically filled with each incoming tide, is now closed and a comprehensive scheme for thoroughly purifying all sewage before discharge is being undertaken. Not only is the sewage now purified, but the sludge produced is added to the island's refuse and processed to produce compost which is used to arrest the declining fertility of the land used for growing early potatoes and tomatoes. Thus a single project which provides for the proper use of organic wastes benefits the island's two main sources of revenue, the tourist trade and market gardening.

Visual evidence of pollution can be concealed if the fragments suspended in the sewage are disintegrated before it is discharged. But this does nothing to provide protection against all the risks of infection which bathing in polluted water carries. To conceal a danger does not remove the risk but merely increases it. Even in the exceptional cases where there is a current flowing constantly out to sea, the suspended matter can never get beyond the continental shelf. There it may help sea-weed to grow, but it does not contribute much to the growth of fish. Our fishermen more than ever have to make hazardous journeys into Arctic waters, there to be exposed to gales and cold as well as the wrath of other nationalities, jealous of the riches contained in their own territorial waters.

It is true that occasionally instances can be cited where the nutrients carried by discharged sewage feed a larger fish population. Off the mouth of the Thames where two great blocks of water from the Atlantic meet, one moving up the English Channel and another round the north of Scotland, fish are caught in twice the numbers they are anywhere else in the North Sea; it is estimated that two-thirds of the excess catch is attributable to the nutrients carried by the Thames. Even so the nutrients contained in the sewage are almost wholly wasted. In phosphorus the increased catch is equivalent to a recovery of about one hundred and twenty tons per year, which is just over 4 per cent of the phosphorus poured into the Thames in sewage: the remaining 96 per cent of this valuable element is wholly lost.

If the discharge of sewage into the sea contributes only occasionally and only to this insignificant amount to the increase of the fish population, it has an unfortunate effect in stimulating the growth of sea-weed. Most of our coastal towns have a problem in sea-weed disposal, even if few are as spectacular or sensational as the regular visitations which fall on the town of Worthing. There, anything up to sixty thousand tons of sea-weed may be thrown up on the foreshore during two months in the summer. At one time farmers were glad to remove these accumulations and use the sea-weed for manuring. They had horses and carts, but the sands will not carry the heavy motor lorries which now replace horse transport and the farmers can moreover use artificial chemical fertilizers instead of the sea-weed, which is left to rot. Holiday-makers are driven from the beaches by the stench of the rotting weed, while an unusual fly of the genus Coelopa breeds in the decaying mass. These flies thrive so well that they blacken the beaches as they rest there in their countless millions. Entomologists from the British Museum identified the flies, but could not suggest anything that would cause their destruction. They were immune to insecticide sprays and cold weather. Attacking them with poison was ruled out because of the danger that would carry to inshore fishing; boarding-house keepers would not allow the sea-weed to be burned in case the fumes would drive their guests from the town. In desperation the Corporation appealed to the Government for help in crushing this threat to the tourist trade, but by the time the Government moved the inconvenience had passed and the problem remains to be solved in some future year.

Because of pollution we may not eat the mussels and winkles that cling to the rocks around our coast. Orders prohibiting the consumption of these shell-fish are common to the local legislation of many coastal authorities. They may be collected and used for bait and they may be exported, ostensibly to be used as bait but in fact to be sold for human consumption, for there is no effective legislation which makes that illegal. And we have virtually written off our oyster-beds — native oysters, once the poor man's food, are now a luxury. A great industry flourishes along the Biscay coast where seven hundred establishments produce thirty thousand tons of oysters a year; the Dutch town of Ierseke with a population of five thousand maintains itself on a basic industry involving the exportation of twenty million oysters, in addition to mussels and crabs, to France and Belgium every year, while we try to persuade a few hundred artificially grown oysters to take kindly to the cold but yet unpolluted waters off the north-west coast of Scotland. The Government regularly pays grants of up to 50 per cent of the cost of works designed to reduce the effects of coastal erosion, but it does very little to discourage coastal authorities from polluting the sea.

Bad as that pollution is, it is trivial compared to the filth of our great river estuaries. The Clyde, the cradle of more ships than any other place in the world, has been more grotesquely altered than any other river in the world. Two hundred years ago a fully grown man could wade across the Clyde ten miles below Glasgow, when its population was a mere twenty-five thousand. Now Glasgow has over one million citizens and because of dredging, which started in 1740, anything but the very largest ocean-going liners can now sail into the heart of the city.

The Clydeside shipbuilders have a skill which is the wonder of the world. They launched the Queen Mary and the Queen Elizabeth into a trout stream, but these streams no longer contain trout. At launchings shipbuilders from every corner of the world come to wonder, and one wondered more than others, for he went away convinced that he had seen a ship of many thousand tons launched into a sewer. St. Mungo, Glasgow's Patron Saint, is reputed to have bathed in the River Kelvin, a tributary of the Clyde, every day throughout the year. If he lived today the health authorities would make him find some other way of doing penance, for although Glasgow now treats all its sewage before discharging it into the Clyde, that only started after its population had risen to three-quarters of a million and the river and its tributaries have not yet recovered from the damage done before that time.

We cannot paint our ships white as is done in Scandinavian countries. After passing once through the Thames estuary every ship comes out black and after one or two passages, the paint comes off altogether. The state of our smaller streams has been, and often is, no better. Two hundred years ago Swift described a stream passing through a town thus: 'Sweepings from butcher stalls, dung, guts and blood, drowned puppies, stinking sprats, all drenched in mud, dead dogs and turnip tops come tumbling down the flood', yet sanitary progress has done little to improve matters since then. Ten years ago H. O. Turing reported to the British Field Sports Society on once lovely rivers rendered 'lifeless, pestiferous, foul, the vomit of our boasted industrial development'. The Lochgelly Burn runs close to the village of Cowdenbeath in the country of Fife. It is not a large stream, and for some obscure reason that may be sound the National Coal Board have the duty to keep it clean. Here is the catalogue of the items taken from it at one spring cleaning: 4 spring mattresses, 26 iron bed ends, 30 angle irons for beds, 6 bolster slips of flock, 16 railway sleepers, 25 pails, 6 big baths, 9 basins, 1 sofa, 1 armchair, 4 household water storage tanks, 4 sinks, 4 wringers, 1 gas stove, 1 old wooden washing machine, 6 clothes poles, 46 old sheets of corrugated iron, 3 prams, 14 smoke boards, 4 fireside fenders, 5 cycle frames, 7 motor tyres, 13 lengths of piping, 1 gas meter, 9 heavy loads of bricks, tin cans and wood, the carcasses of 2 dogs, 1 cat and 4 hens, and the Lochgelly Burn is polluted with sewage to boot.

We are not alone in our difficulties. The Catholic Farmers Union of Canada concerned with the need for conserving the natural resources of the country, have asked the Government to form a comprehensive public policy towards land, forest and water. The Government already has a commission studying the pollution of rivers which was established after it was known that all the water flowing around the island of Montreal was so poisoned that it was considered unfit for bathing, now they are being urged to set up a new ministry to deal with all conservation problems. The Swiss are also in trouble with the sewage they discharge into their great lakes. The volume of water in these is so enormous, that up until the present time the lower depths have provided good water supplies while crude sewage was discharged near the surface. But as long ago as 1825 the surface waters of the small lake of Murten turned red overnight. Legend had it that this was the blood of Burgundian warriors slain by Swiss patriots in battle in 1476. It is now known that it was due to the rapid growth of a small water weed which was nourished by salts released from unpurified sewage. Since 1825 the growth of weed has spread to the bigger lakes. In 1952 Lake Lugano was affected and now it has spread to high-level reservoirs.

Other similar examples could be quoted in this country. They are by no means rare. It is astonishing that we have not yet had the wits to put an end to these disgusting and intolerable conditions. We have had time. It is nearly one hundred years since Members of Parliament found the stench from the Thames so nauseating that they could not use some of their committee rooms and were consequently moved to appoint a committee to find a remedy for river pollution. The first committee of inquiry was appointed in 1862 and was followed by others, but the only important and effective one was the Royal Commission on Sewage Disposal which was appointed in 1898. This Commission sat for seventeen dreary years before being dissolved, when it published its tenth and final report. Not all its recommendations, including what are probably its more important ones covering river pollution, have been incorporated in statutes, but its findings have had a great influence on local health authorities, many of whom have freely accepted a number of standards recommended by the Commissioners.

It is essential to understand the natural process by means of which running streams have the capacity, while carrying off sewage, to purify it; it was on this process that the findings of the Commission were based and they have remained operative ever since. The actual purifying agent is oxygen, which will combine with the noxious sewage matter and render it harmless. There are two sources of oxygen, namely, the oxygen dissolved in the stream waters themselves and the oxygen in the air; a stream as it flows along, has the capacity to take up atmospheric oxygen and the more turbulent is its flow the greater is the amount of the additional oxygen it can absorb in this way. This is a very important fact.

It is clear therefore that the capacity of a stream to deal effectively with sewage depends both on its volume and its rate of flow. A larger stream is more effective than a smaller one: a fast-flowing stream more effective than a slow one of the same size. But if either the volume or the strength of the sewage which is added to the stream absorbs more oxygen than the stream can supply or can take up from the atmosphere, the purification process will be incomplete and the stream will remain polluted because it will still be carrying unoxidized sewage.

The immediate arbiters of river pollution are fish, especially those of the finer species. They require oxygen to keep alive and if the sewage liquor absorbs a high proportion of the available oxygen from the river, a point may be reached where the stream cannot oxidize the sewage and sustain a fish population at the same time, and the fish die off. Contrary to popular belief the poisoning of fish by actual toxic substances contained in sewage is a much less common reason for the reduction of the fish population than their asphyxiation through lack of oxygen.

The oxygen content of water was consequently adopted by the Commission as the principal factor by which the degree of ultimate purification or of pollution could be determined and, with a view to ensuring that the oxygen content of streams would not be reduced below the point which would allow fish to thrive, suggested that before crude, or treated, sewage was discharged into streams it should first be tested to establish how much oxygen it would absorb.

The oxygen required to purify sewage liquor can be readily established and is known as the 'biochemical oxygen demand' or B.O.D. of the liquor. When this has been determined and the total volume of sewage to be discharged into a stream is known, it can be related to the oxygen which is available from the stream and to the stream's power to recover oxygen from the atmosphere.

The object of sewage treatment is therefore to reduce the biochemical oxygen demand of sewage so as not to overtax the stream, and the Commission suggested that if this demand were first determined and then related to the oxygen available from the stream which is to receive it, the amount of artificial treatment the sewage required could be established. As a further precaution against pollution it also recommended that the amount of organic matter suspended should be determined and related to the volume of water flowing in the stream in dry weather, and they gave guidance on the determination of the treatment required under varying circumstances.

The work of the Commission was fundamentally useful. We are supplied with abundant safe water for drinking and other purposes. The public has reason to be grateful; the Members of Parliament can enjoy tea on their terrace overlooking the Thames; the members of the area boards appointed under the Rivers Pollution Act of 1951 have the power and the means to do their work properly, but careful examination reveals a far from ideal state of affairs. Those other arbiters, the fish, are less easily satisfied. The Thames and the Clyde and the Severn and other great waterways of Britain were, in comparatively recent times, gateways through which salmon and sea trout in their countless thousands passed into the streams of which Britain is so plentiful to provide food and good sport. Forty years ago twenty-five thousand salmon were taken out of the Severn. Now the numbers have dropped to between three and four thousand. The corresponding figures for the Tyne are one hundred and twenty thousand against seven hundred today and for the Tees eight thousand which had dropped to nil and the Clyde and the Thames are almost deserted. And in 1953 we spent one and a half million pounds in fabulous dollars to bring salmon in tins from North America.

When salmon were so plentiful that they could be taken from almost any stream in Britain with a pitchfork, the better-off found them too cloying to be enjoyed and the working class would not have them served at their tables oftener than twice a week. Today we can get salmon coloured pink out of a tin, or we can have it smoked when it is as exotic as caviare and almost as expensive. Thanks to river pollution salmon now have a scarcity value that works out at about £5 a fish and gangsters, as tough as anything produced by prohibition in America, blast the remaining persecuted salmon from their rivers with high explosives and sell what they catch in town markets, while landlords, who get huge rents for the few stretches of good fishing waters that remain, wage a war to the death with the gangster poachers from the cities.

The waters of some streams in the north of Scotland are now being controlled to produce electrical energy to take some pressure off the demands put upon our diminishing coal stocks. These are amongst the few remaining waters in Britain which carry salmon and trout. When they are being dammed, complicated provisions are being made to allow the fish to pass freely from the sea to the spawning grounds in the head waters and to return to the sea again. Dams over one hundred feet in height are being constructed within which fish passes and lifts are being built which work out at a cost of about two thousand pounds per foot rise. Thus we have a situation where a public authority at one end of a river may be killing fish through pollution, while at the other end another public authority is spending perhaps a quarter of a million pounds to keep alive the few fish that escape.

What fish there is outside salmon rivers is coarse. Compare the pike, living interminably and growing enormous on the easy feeding which decaying wastes provide, with the noble salmon or courageous trout who refuse to live amongst filth. Pike can be lured on to a hook by a lump of agenized bread or a piece of red flannel and once hooked will show such limited desire to continue life in their miserable surroundings that they can be pulled on to the bank with no more show of resistance than that offered by their own gross weight. The network of streams in this country are vital to our continued progress, but even birds desert these desecrated waters and once gone are not easily persuaded to return. But man with his domesticated animals must remain, while he continues with the debasement of this essential element in an environment which made him human.

The results are depressing because an enormous expenditure of money, effort, and skill have gone into building up our present system. In spite of the defects described we have much to be proud of. Today sewage, measured in gallons by the hundred million, is daily carried by the sewers of towns and villages into rivers or the sea, sometimes after elaborate treatment. This enormous discharge is directly related to an enormous intake of water for various uses; the problems of intake of pure water at one end and its discharge when soiled at another are really one and the same problem and have to be solved at either end by not dissimilar mechanical devices. These are on a vast scale and are complicated. More than 90,000 miles of pipes, aqueducts and tunnels, the vast majority of them laid in the last eighty years, bring water from hills, lakes, rivers and wells to houses and factories and additional miles are being laid every day. The forest of poles which everyone can see above ground carrying public lights, telegraph wires, power, and decorative things like flags and bunting and such mundane matters as washings is as nothing compared with the hidden labyrinth of pipes carrying sewage, water, gas, electric and post office cables which exist under every city street. A tunnel six feet in diameter — a not unusual size — bringing clean water into a city today costs £100 a yard to construct and the three-inch pipe which is required to bring such water in the quantity necessary to meet the demands of the smallest group of houses costs thirty shillings a yard by the time it is in the ground. The reservoirs, deep wells, storage tanks, valves and pumping stations, with their complex fittings which control the rate of flow and ensure that everyone gets water at the right pressure, built when labour was cheap, are now beyond price. For no longer can we stop and drink at a stream in this crowded island without thought for the consequences; a grossly polluted atmosphere and countryside now make that impossible.

Yet there is no shortage of water. In the dry south of England about twenty inches of rain fall every year, while some parts of the Lake District and north-west of Scotland would be submerged to a depth of ten feet if the rain that falls during a year did not gravitate to the sea. Water in adequate quantity is available to everyone in this country merely for the trouble it takes to collect it; it is clean water that is scarce and costly. More than two thousand million gallons of water are used by the people of London each year. Four-fifths of this supply comes from the Thames and the Lee and only the perfection of sewage treatment processes makes that possible, for much of the water used by Londoners has already been used by populations higher up these rivers. The most careful control is required to allow these waters to be used with complete safety to augment the supply of clean hard water drawn from the London chalk. The skill acquired in the treatment of sewage makes it unnecessary for London to go the three hundred miles to the Lake District for its water. Other towns are more favourably situated. Manchester now makes use of these lakes but Glasgow is even better placed, finding the millions of gallons it consumes daily in the top few feet of nearby Loch Katrine.

The water which is thus brought into the great cities (after treatment in filters, and chemicals added with as much care as is taken in a hospital laboratory) is used at the rate of between thirty and sixty gallons per head of the population per day, and the demand grows. After use, it is discharged as sewage into a reticulation of pipes, which if placed end to end would gird the earth four times at the Equator and is as complex in its working as the system which brings the clean water in. Appliances as elaborate as those required for the distribution of water, are needed to take the used water away, and men have to go into the sewers to keep them running freely.

The miner who spends the whole of his working life in the depths of the earth hewing coal for industry and the home, has won for himself an honoured status in our society and receives respect and great sympathy for the sacrifices he makes for our well being and comfort; but the devotion of the sewermen, who work under conditions which are infinitely worse than any miner, passes unnoticed, possibly because they are fewer in number and have no trade union of their own. Yet were they to fail us, the consequences would be more damaging and far-reaching than all the discomforts and inconveniences of power cuts. For eight hours a day and five days a week these men work in the darkness, wading up to the waist in sewage, with sewer rats as their only living companions. They remove debris so that we who find sewage too nauseating to contemplate may enjoy all the advantages of press-button sanitation. It is the efficiency of maligned British plumbing and of these men that we have to thank for the fact that sewage welling up in gardens or thoroughfares is such a rare occurrence that it warrants mention in the front page of national papers when it happens.

But the health of the sewermen is good and the extra pound a week their odious work brings makes the job much sought after by those engaged on this essential service. Even then their pay falls far short of the miners' and if the miner is constantly threatened by explosive fire damp, the sewerman's job carries its own dangers. On 10th September 1953, the trunk sewer in one of the main streets of a large American city exploded and the blast ripped the street and pavements wide upon for a distance of over a mile. Miraculously, only one person was killed, although sixty-four were taken to hospital, and how many more suffered minor injuries and shock is not known. The damage to property cost one and a half million dollars to repair, two-thirds of which was for sewers and paving and the rest for restoring other damaged services. Damage to private property and intangible costs were estimated at another million dollars. It was an unusual occurrence, but the cause of the disaster was finally established as being due to an explosive mixture of nothing more unusual than industrial wastes and petrol which had seeped into the sewer through a fault. The diligence of the sewermen and their fellow workers keep our cities and towns clean. They have removed the curse of dysentery, cholera and typhoid and no blame can be attached to them that the sewage which they direct away from our towns is destined to pollute rivers and bathing beaches instead of fertilizing the land.

The huge volumes of sewage which daily discharge from house and factory exceed in total the flow in the largest rivers in the country. Indeed, the effluent from the recently constructed sewage purification works at Mogden, which serves the population of West Middlesex, is greater in quantity in dry weather than the flow in the Thames at the point at which it is discharged and it is often cleaner. Despite the magnitude of the task the power to stop river pollution is available to us, but unless we make greater use of new processes for treating sewage sludge, improvements in sewage purification cannot be fully exploited and polluted rivers and coastal waters will continue to be suffered.

The Grapes of Wrath *

John Steinbeck

Chapter One

To the red country and part of the gray country of Oklahoma, the last rains came gently, and they did not cut the scarred earth. The plows crossed and recrossed the rivulet marks. The last rains lifted the corn quickly and scattered weed colonies and grass along the sides of the roads so that the gray country and the dark red country began to disappear under a green cover. In the last part of May the sky grew pale and the clouds that had hung in high puffs for so long in the spring were dissipated. The sun flared down on the growing corn day after day until a line of brown spread along the edge of each green bayonet. The clouds appeared, and went away, and in a while they did not try any more. The weeds grew darker green to protect themselves, and they did not spread any more. The surface of the earth crusted, a thin hard crust, and as the sky became pale, so the earth became pale, pink in the red country and white in the gray country.

In the water-cut gullies the earth dusted down in dry little streams. Gophers and ant lions started small avalanches. And as the sharp sun struck day after day, the leaves of the young corn became less stiff and erect; they bent in a curve at first, and then, as the central ribs of strength grew weak, each leaf tilted downward. Then it was June, and the sun shone more fiercely. The brown lines of the corn leaves widened and moved in on the central ribs. The weeds frayed and edged back toward their roots. The air was thin and the sky more pale; and every day the earth paled.

In the roads where the teams moved, where the wheels milled the ground and the hooves of the horses beat the ground, the dirt crust broke and the dust formed. Every moving thing lifted the dust into the air: a walking man lifted a thin layer as high as his waist, and a wagon lifted the dust as high as the fence

---

*From The Grapes of Wrath by John Steinbeck. Copyright 1939, copyright ©renewed 1967 by John Steinbeck Reprinted by permission of The Viking Press, Inc.

tops, and an automobile boiled a cloud behind it. The dust was long in settling back again.

When June was half gone, the big clouds moved up out of Texas and the Gulf, high heavy clouds, rain-heads. The men in the fields looked up at the clouds and sniffed at them and held wet fingers up to sense the wind. And the horses were nervous while the clouds were up. The rain-heads dropped a little spattering and hurried on to some other country. Behind them the sky was pale again and the sun flared. In the dust there were drop craters where the rain had fallen, and there were clean splashes on the corn, and that was all.

A gentle wind followed the rain clouds, driving them on northward, a wind that softly clashed the drying corn. A day went by and the wind increased, steady, unbroken by gusts. The dust from the roads fluffed up and spread out and fell on the weeds beside the fields, and fell into the fields a little way. Now the wind grew strong and hard and it worked at the rain crust in the corn fields. Little by little the sky was darkened by the mixing dust, and the wind felt over the earth, loosened the dust, and carried it away. The wind grew stronger. The rain crust broke and the dust lifted up out of the fields and drove gray plumes into the air like sluggish smoke. The corn threshed the wind and made a dry, rushing sound. The finest dust did not settle back to earth now, but disappeared into the darkening sky.

The wind grew stronger, whisked under stones, carried up straws and old leaves, and even little clods, marking its course as it sailed across the fields. The air and the sky darkened and through them the sun shone redly, and there was a raw sting in the air. During a night the wind raced faster over the land, dug cunningly among the rootlets of the corn, and the corn fought the wind with its weakened leaves until the roots were freed by the prying wind and then each stalk settled wearily sideways toward the earth and pointed the direction of the wind.

The dawn came, but no day. In the gray sky a red sun appeared, a dim red circle that gave a little light, like dusk; and as that day advanced, the dusk slipped back toward darkness, and the wind cried and whimpered over the fallen corn.

Men and women huddled in their houses, and they tied handkerchiefs over their noses when they went out, and wore goggles to protect their eyes.

When the night came again it was black night, for the stars could not pierce the dust to get down, and the window lights could not even spread beyond their own yards. Now the dust was evenly mixed with the air, an emulsion of dust and air. Houses were shut tight, and cloth wedged around doors and windows, but the dust came in so thinly that it could not be seen in the air, and it settled like pollen on the chairs and tables, on the dishes. The people brushed it from their shoulders. Little lines of dust lay at the door sills.

In the middle of that night the wind passed on and left the land quiet. The dust-filled air muffled sound more completely than fog does. The people, lying in their beds, heard the wind stop. They awakened when the rushing wind was gone. They lay quietly and listened deep into the stillness. Then the roosters crowed, and their voices were muffled, and the people stirred restlessly in their beds and wanted the morning. They knew it would take a long time for the dust to settle out of the air. In the morning the dust hung like fog, and the sun was as red as ripe new blood. All day the dust sifted down from the sky, and the next day it sifted down. An even blanket covered the earth. It settled on the corn, piled up on the tops of the fence posts, piled up on the wires; it settled on roofs, blanketed the weeds and trees.

The people came out of their houses and smelled the hot stinging air and covered their noses from it. And the children came out of the houses, but they did not run or shout as they would have done after a rain. Men stood by their fences and looked at the ruined corn, drying fast now, only a little green showing through the film of dust. The men were silent and they did not move often. And the women came out of the houses to stand beside their men — to feel whether this time the men would break. The women studied the men's faces secretly, for the corn could go, as long as something else remained. The children stood near by, drawing figures in the dust with bare toes, and the children sent exploring senses out to see whether men and women would break. The children peeked at the faces of the men and women, and then drew careful lines in the dust with their toes. Horses came to the watering troughs and nuzzled the water to clear the surface dust. After a while the faces of the watching men lost their bemused perplexity and became hard and angry and resistant. Then the women knew that they were safe and that there was no break. Then they asked, What'll we do? And the men replied, I don't know. But it was all right. The women knew it was all right, and the watching children knew it was all right. Women and children knew deep in themselves that no misfortune was too great to bear if their men were whole. The women went into the houses to their work, and the children began to play, but cautiously at first. As the day went forward the sun became less red. It flared down on the dust-blanketed land. The men sat in the doorways of their houses; their hands were busy with sticks and little rocks. The men sat still — thinking — figuring.

Chapter Five

The owners of the land came onto the land, or more often a spokesman for the owners came. They came in closed cars, and they felt the dry earth with their fingers, and sometimes they drove big earth augers into the ground for soil tests. The tenants, from their sun-beaten dooryards, watched uneasily when the closed cars drove along the fields. And at last the owner men drove into the dooryards and sat in their cars to talk out of the windows. The tenant men stood beside the cars for a while, and then squatted on their hams and found sticks with which to mark the dust.

In the open doors the women stood looking out, and behind them the children — corn-headed children, with wide eyes, one bare foot on top of the other bare foot, and the toes working. The women and the children watched their men talking to the owner men. They were silent.

Some of the owner men were kind because they hated what they had to do, and some of them were angry because they hated to be cruel, and some of them were cold because they had long ago found that one could not be an owner unless one were cold. And all of them were caught in something larger than themselves. Some of them hated the mathematics that drove them, and some were afraid, and some worshiped the mathematics because it provided a refuge from thought and from feeling. If a bank or a finance company owned the land, the owner man said, The Bank — or the Company — needs — wants — insists — must have — as though the Bank or the Company were a monster, with thought and feeling, which had ensnared them. These last would take no responsibility for the banks or the companies because they were men and slaves, while the banks were machines and masters all at the same time. Some of the owner men were a little proud to be slaves to such cold and powerful masters. The owner men sat in the cars and explained. You know the land is poor. You've scrabbled at it long enough, God knows.

The squatting tenant men nodded and wondered and drew figures in the dust, and yes, they knew, God knows. If the dust only wouldn't fly. If the top would only stay on the soil, it might not be so bad.

The owner men went on leading to their point: You know the land's getting poorer. You know what cotton does to the land; robs it, sucks all the blood out of it.

The squatters nodded — they knew, God knew. If they could only rotate the crops they might pump blood back into the land.

Well, it's too late. And the owner men explained the workings and the thinkings of the monster that was stronger than they were. A man can hold land if he can just eat and pay taxes; he can do that.

Yes, he can do that until his crops fail one day and he has to borrow money from the bank.

But — you see, a bank or a company can't do that, because those creatures don't breathe air, don't eat side-meat. They breathe profits; they eat the interest on money. If they don't get it, they die the way you die without air, without side-meat. It is a sad thing, but it is so. It is just so.

The squatting men raised their eyes to understand. Can't we just hang on? Maybe the next year will be a good year. God knows how much cotton next year. And with all the wars — God knows what price cotton will bring. Don't they make explosives out of cotton? And uniforms? Get enough wars and cotton'll hit the ceiling. Next year, maybe. They looked up questioningly.

We can't depend on it. The bank — the monster has to have profits all the time. It can't wait. It'll die. No, taxes go on. When the monster stops growing, it dies. It can't stay one size.

Soft fingers began to tap the sill of the car window, and hard fingers tightened on the restless drawing sticks. In the doorways of the sun-beaten tenant houses, women sighed and then shifted feet so that the one that had been down was now on top, and the toes working. Dogs came sniffing near the owner cars and wetted on all four tires one after another. And chickens lay in the sunny dust and fluffed their feathers to get the cleansing dust down to the skin. In the little sties the pigs grunted inquiringly over the muddy remnants of the slops.

The squatting men looked down again. What do you want us to do? We can't take less share of the crop — we're half starved now. The kids are hungry all the time. We got no clothes, torn an' ragged. If all the neighbors weren't the same, we'd be ashamed to go to meeting.

And at last the owner men came to the point. The tenant system won't work any more. One man on a tractor can take the place of twelve or fourteen families. Pay him a wage and take all the crop. We have to do it. We don't like to do it. But the monster's sick. Something's happened to the monster.

But you'll kill the land with cotton.

We know. We've got to take cotton quick before the land dies. Then we'll sell the land. Lots of families in the East would like to own a piece of land.

The tenant men looked up alarmed. But what'll happen to us? How'll we eat?

You'll have to get off the land. The plows'll go through the dooryard.

And now the squatting men stood up angrily. Grampa took up the land, and he had to kill the Indians and drive them away. And Pa was born here, and he killed weeds and snakes. Then a bad year came and he had to borrow a little money. An' we was born here. There in the door — our children born here. And Pa had to borrow money. The bank owned the land then, but we stayed and we got a little bit of what we raised.

We know that — all that. It's not us, it's the bank. A bank isn't like a man. Or an owner with fifty thousand acres, he isn't like a man either. That's the monster.

Sure, cried the tenant men, but it's our land. We measured it and broke it up. We were born on it, and we got killed on it, died on it. Even if it's no good, it's still ours. That's what makes it ours — being born on it, working it, dying on it. That makes ownership, not a paper with numbers on it.

We're sorry. It's not us. It's the monster. The bank isn't like a man.

Yes, but the bank is only made of men.

No, you're wrong there — quite wrong there. The bank is something else than men. It happens that every man in a bank hates what the bank does, and yet the bank does it. The bank is something more than men, I tell you. It's the monster. Men made it, but they can't control it.

The tenants cried, Grampa killed Indians, Pa killed snakes for the land. Maybe we can kill banks — they're worse than Indians and snakes. Maybe we got to fight to keep our land, like Pa and Grampa did.

And now the owner men grew angry. You'll have to go.

But it's ours, the tenant men cried. We —

No. The bank, the monster owns it. You'll have to go.

We'll get our guns, like Grampa when the Indians came. What then?

Well — first the sheriff, and then the troops. You'll be stealing if you try to stay, you'll be murderers if you kill to stay. The monster isn't men, but it can make men do what it wants.

But if we go, where'll we go? How'll we go? We got no money.

We're sorry, said the owner men. The bank, the fifty-thousand-acre owner can't be responsible. You're on land that isn't yours. Once over the line maybe you can pick cotton in the fall. Maybe you can go on relief. Why don't you go on west to California? There's work there, and it never gets cold. Why, you can reach out anywhere and pick an orange. Why, there's always some kind of crop to work in. Why don't you go there? And the owner men started their cars and rolled away.

The tenant men squatted down on their hams again to mark the dust with a stick, to figure, to wonder. Their sunburned faces were dark, and their sun-whipped eyes were light. The women moved cautiously out of the doorways toward their men, and the children crept behind the women, cautiously, ready to run. The bigger boys squatted beside their fathers, because that made them men. After a time the women asked, What did he want?

And the men looked up for a second, and the smolder of pain was in their eyes. We got to get off. A tractor and a superintendent. Like factories.

Where'll we go? the women asked.

We don't know. We don't know.

And the women went quickly, quietly back into the houses and herded the children ahead of them. They knew that a man so hurt and so perplexed may turn in anger, even on people he loves. They left the men alone to figure and to wonder in the dust.

After a time perhaps the tenant man looked about — at the pump put in ten years ago, with a goose-neck handle and iron flowers on the spout, at the chopping block where a thousand chickens had been killed, at the land plow lying in the shed, and the patent crib hanging in the rafters over it.

The children crowded about the women in the houses. What we going to do, Ma? Where we going to go?

The women said, We don't know, yet. Go out and play. But don't go near your father. He might whale you if you go near him. And the women went on with the work, but all the time they watched the men squatting in the dust — perplexed and figuring.

The tractors came over the roads and into the fields, great crawlers moving like insects, having the incredible strength of insects. They crawled over the ground, laying the track and rolling on it and picking it up. Diesel tractors, puttering while they stood idle; they thundered when they moved, and then settled down to a droning roar. Snub-nosed monsters, raising the dust and sticking their snouts into it, straight down the country, across the country, through fences, through dooryards, in and out of gullies in straight lines. They did not run on the ground, but on their own roadbeds. They ignored hills and gulches, water courses, fences, houses.

The man sitting in the iron seat did not look like a man; gloved, goggled, rubber dust mask over nose and mouth, he was a part of the monster, a robot in the seat. The thunder of the cylinders sounded through the country, became one with the air and the earth, so that earth and air muttered in sympathetic vibration. The driver could not control it — straight across country it went, cutting through a dozen farms and straight back. A twitch at the controls could swerve the cat', but the driver's hands could not twitch because the monster that built the tractor, the monster that sent the tractor out, had somehow got into the driver's hands, into his brain and muscle, had goggled him and muzzled him — goggled his mind, muzzled his speech, goggled his perception, muzzled his protest. He could not see the land as it was, he could not smell the land as it smelled; his feet did not stamp the clods or feel the warmth and power of the earth. He sat in an iron seat and stepped on iron pedals. He could not cheer or beat or curse or encourage the extension of his power, and because of this he could not cheer or whip or curse or encourage himself. He did not know or own or trust or beseech the land. If a seed dropped did not germinate, it was nothing. If the young thrusting plant withered in drought or drowned in a flood of rain, it was no more to the driver than to the tractor.

He loved the land no more than the bank loved the land. He could admire the tractor — its machined surfaces, its surge of power, the roar of its detonating cylinders; but it was not his tractor. Behind the tractor rolled the shining disks, cutting the earth with blades — not plowing but surgery, pushing the cut earth to the right where the second row of disks cut it and pushed it to the left; slicing blades shining, polished by the cut earth. And pulled behind the disks, the harrows combing with iron teeth so that the little clods broke up and the earth lay smooth. Behind the harrows, the long seeders — twelve curved iron penes erected in the foundry, orgasms set by gears, raping methodically, raping without passion. The driver sat in his iron seat and he was proud of the straight lines he did not will, proud of the tractor he did not own or love, proud of the power he could not control. And when that crop grew, and was harvested, no man had crumbled a hot clod in his fingers and let the earth sift past his fingertips. No man had touched the seed or lusted for the growth. Men ate what they had not raised, had no connection with the bread. The land bore under iron, and under iron gradually died; for it was not loved or hated, it had no prayers or curses.

At noon the tractor driver stopped sometimes near a tenant house and opened his lunch: sandwiches wrapped in waxed paper, white bread, pickle, cheese, Spam, a piece of pie branded like an engine part. He ate without relish. And tenants not yet moved away came out to see him, looked curiously while the goggles were taken off, and the rubber dust mask, leaving white circles around the eyes and a large white circle around nose and mouth. The exhaust of the tractor puttered on, for fuel is so cheap it is more efficient to leave the engine running than to heat the Diesel nose for a new start. Curious children crowded close, ragged children who ate their fried dough as they watched. They watched hungrily the unwrapping of the sandwiches, and their hunger-sharpened noses smelled the pickle, cheese, and Spam. They didn't speak to the driver. They watched his hand as it carried food to his mouth. They did not watch him chewing; their eyes followed the hand that held the sandwich. After a while the tenant who could not leave the place came out and squatted in the shade beside the tractor.

"Why, you're Joe Davis's boy!"

"Sure," the driver said.

"Well, what you doing this kind of work for — against your own people?"

"Three dollars a day. I got damn sick of creeping for my dinner — and not getting it. I got a wife and kids. We got to eat. Three dollars a day, and it comes every day."

"That's right," the tenant said. "But for your three dollars a day fifteen or twenty families can't eat at all. Nearly a hundred people have to go out and wander on the roads for your three dollars a day. Is that right?"

And the driver said, "Can't think of that. Got to think of my own kids. Three dollars a day, and it comes every day. Times are changing, mister, don't you know? Can't make a living on the land unless you've got two, five, ten thousand acres and a tractor. Crop land isn't for little guys like us any more. You don't kick up a howl because you can't make Fords, or because you're not the telephone company. Well, crops are like that now. Nothing to do about it. You try to get three dollars a day someplace. That's the only way."

The tenant pondered. "Funny thing how it is. If a man owns a little property, that property is him, it's part of him, and it's like him. If he owns property only so he can walk on it and handle it and be sad when it isn't doing well, and feel fine when the rain falls on it, that property is him, and some way he's bigger because he owns it. Even if he isn't successful he's big with his property. That is so."

And the tenant pondered more. "But let a man get property he doesn't see, or can't take time to get his fingers in, or can't be there to walk on it — why, then the property is the man. He can't do what he wants, he can't think what he wants. The property is the man, stronger than he is. And he is small, not big. Only his possessions are big — and he's the servant of his property. That is so, too."

The driver munched the branded pie and threw the crust away. "Times are changed, don't you know? Thinking about stuff like that don't feed the kids. Get your three dollars a day, feed your kids. You got no call to worry about anybody's kids but your own. You get a reputation for talking like that, and you'll never get three dollars a day. Big shots won't give you three dollars a day if you worry about anything but your three dollars a day."

"Nearly a hundred people on the road for your three dollars. Where will we go?"

"And that reminds me," the driver said, "you better get out soon. I'm going through the dooryard after dinner."

"You filled in the well this morning."

"I know. Had to keep the line straight. But I'm going through the dooryard after dinner. Got to keep the lines straight. And — well, you know Joe Davis, my old man, so I'll tell you this. I got orders wherever there's a family not moved out — if I have an accident — you know, get too close and cave the house in a little — well, I might get a couple of dollars. And my youngest kid never had no shoes yet."

"I built it with my hands. Straightened old nails to put the sheathing on. Rafters are wired to the stringers with baling wire. It's mine. I built it. You bump it down — I'll be in the window with a rifle. You even come too close and I'll pot you like a rabbit."

"It's not me. There's nothing I can do. I'll lose my job if I don't do it. And look — suppose you kill me? They'll just hang you, but long before you're hung there'll be another guy on the tractor, and he'll bump the house down. You're not killing the right guy."

"That's so," the tenant said. "Who gave you orders? I'll go after him. He's the one to kill."

"You're wrong. He got his orders from the bank. The bank told him, 'Clear those people out or it's your job.' "

"Well, there's a president of the bank. There's a board of directors. I'll fill up the magazine of the rifle and go into the bank."

The driver said, "Fellow was telling me the bank gets orders from the East. The orders were, 'Make the land show profit or we'll close you up.' "

"But where does it stop? Who can we shoot? I don't aim to starve to death before I kill the man that's starving me."

"I don't know. Maybe there's nobody to shoot. Maybe the thing isn't men at all. Maybe, like you said, the property's doing it. Anyway I told you my orders."

"I got to figure," the tenant said. "We all got to figure. There's some way to stop this. It's not like lightening or earthquakes. We've got a bad thing made by men, and by God that's something we can change." The tenant sat in his doorway, and the driver thundered his engine and started off, tracks falling and curving, harrows combing, and the phalli of the seeder slipping into the ground. Across the dooryard the tractor cut, and the hard, foot-beaten ground was seeded field, and the tractor cut through again; the uncut space was ten feet wide. And back he came. The iron guard bit into the house-corner, crumbled the wall, and wrenched the little house from its foundation so that it fell sideways, crushed like a bug. And the driver was goggled and a rubber mask covered his nose and mouth. The tractor cut a straight line on, and the air and the ground vibrated with its thunder. The tenant man stared after it, his rifle in his hand. His wife was beside him, and the quiet children behind. And all of them stared after the tractor.

Pollution of the Environment:
An Essay in Six Parts

1. The automobile:
exhaust fumes and other debris

The most effective means by which the population of the United States could be exposed to a particular substance would be to add it to gasoline. Had the proponents of fluoridation realized this, the long battle over the addition of fluorides to public drinking water might have been avoided, for the gasoline pump is now a better dispenser than the water faucet.

During the past year the United States consumed 80 billion gallons of gasoline. Each gallon used produces roughly three pounds of carbon monoxide and two ounces of mixed oxides of nitrogen; the total production of these unwanted substances was 120 million tons of carbon monoxide and 5 million tons of oxides of nitrogen. These are gaseous wastes that are added to the air we breathe. Because gasoline contains tetraethyl lead, an antiknock agent, the exhaust fumes also contain lead (in the form of lead bromide); about one-quarter billion tons of lead are given off by automobile and truck traffic each year. Dr. Joshua Lederberg* has recently called attention to the newly adopted use of nickel as a gasoline additive that, among other things, reduces deposits in the engine; presumably this metal leaves the car by way of the exhaust as nickel carbonyl. Furthermore, those substances that would otherwise have been engine deposits become, thanks to the use of nickel, additional air pollutants.

The purpose of this essay is to point out some of the biological properties of the waste products of automobile traffic. I shall touch on the waste products not only of the exhaust system but also on those of the brake linings and tires: asbestos fibers and particles of vulcanized rubber.

*Joshua Lederberg is a Nobel laureate in Medicine. Born in 1925, he was by the age of 21 the co-discoverer of the transfer of chromosomal material between conjugating bacterial cells. Because of his brilliant work in microbial genetics, Dr. Lederberg was elected to the National Academy of Sciences in 1957 and, in 1958, was awarded the Nobel Prize together with George W. Beadle and Edward L. Tatum. In recent years Dr. Lederberg has been increasingly active in matters of environmental affairs and of the social consequences of scientific endeavors.

143

Many city dwellers have detectable quantities of asbestos fibers in their lungs; brake linings are an important source of these fibers. Experience with industrial workers exposed to asbestos dust has shown that asbestos is associated with lung cancer and other diseases; little else is known about its effect on health.

Nearly 100 million automobiles (including trucks and busses) are registered in the United States. If we allow for the extra wheels on larger vehicles, one-half billion is a reasonable estimate of the number of tires in use at any moment. At an average weight of 40 pounds per tire, a half-billion tires represents 20 billion pounds of vulcanized rubber; of this, 2 to 3 billion pounds are worn off annually in traffic. This rubber, together with the industrial additives necessary for its manufacture, is thrown off as small particles. Among the additives one encounters in rubber are sulfur (100 million pounds in the worn-off particles alone) and epoxy compounds (among the most potent mutagenic substances known to geneticists).

Lead compounds have insidious effects on man; lead poisoning and the danger of using lead cooking utensils and lead water pipes have been know for centuries. An estimated one-quarter million American children fall victim to lead poisoning each year by eating flakes of lead paint and window putty. The amount of lead that accumulates in persons today largely depends on their place of residence and occupational exposure to exhaust fumes; the amount of lead encountered in general, widespread surveys is not sufficient to cause alarm. The recognized safe dose of lead is one-half milligram per day; an accumulated level of 80 milligrams in the blood stream causes damage, at times irreparable, to the brain and nervous system. The "safe" dose is one calculated for clinical purposes however; a "safe" dose for the exposure of an entire population from birth to death has not been established.

By 1975 the estimated automobile traffic in the United States will be double the present level; one-quarter of this traffic will travel 40,000 miles of the interstate highway system. At that level of traffic, an average of 3,000 pounds of lead will be spewed out per mile of highway per year. This level will be much higher at certain focal points where traffic funnels into large cities and, conversely, will be much lower in the less traveled rural sections of the country. There is serious question whether the present safe dose will be an adequate guideline for the exposure of city children raised from birth along the arterial highways of Boston, New York, Philadelphia, Pittsburgh, Chicago, Baltimore, Detroit, Los Angeles, San Francisco, Seattle, or any other large city. Plants that live today near existing superhighways are frequently lead-resistant; these plants are able to live at roadsides solely by virtue of evolutionary changes comparable to those that are responsible for DDT-resistance in insects and antibiotic-resistance bacteria.

The danger of nickel as a newly found pollutant has been emphasized by Dr. Joshua Lederberg. Studies on both mice and nickel refinery workers have shown that nickel compounds (especially the one likely to be formed by burning nickel-containing gasoline) cause cancer of the nasal passages and of the lungs. The addition of this compound to gasoline is but one more indication of the general myopia of industrial leaders and industrial engineers. "That greed," says Dr. Lederberg, "responsive to narrow goals and blind to larger human needs, can have no end other than a terminal cosmic bellyache."* The word Dr. Lederberg chose is "terminal," not "giant" or "colossal" or "worrisome." If greed and lack of foresight continue unabated on the scale of which man is now capable of performing, our descendants will literally pay with their lives.

Carbon monoxide is the colorless and odorless gas in auto exhaust that kills those who run their engines in closed garages or whose cars have faulty exhaust systems. The affinity of carbon monoxide for hemoglobin is nearly three hundred times as great as that of oxygen, and so a mere trace of carbon monoxide in the air we breathe can effectively destory the oxygen-transporting ability of the blood. Cirgarette smokers lose about 5 per cent of their hemoglobin to the stable hemoglobin-carbon monoxide combination; city dwellers exposed to 20 parts carbon monoxide in 1 million parts of air have about 3 per cent of their hemoglobin blocked. Once more, the physiological effects of chronic exposure to carbon monoxide are virtually unknown — especially during childhood years when growth proceeds rapidly, during pregnancies when the fetus makes tremendous demands upon the mother's circulation, and during old age and debilitating illnesses when the body is especially sensitive to extra stresses.

Automotive exhausts are not the only sources of carbon monoxide in the atmosphere; about one-quarter billion tons are poured into the atmosphere annually by worldwide industry. Despite the rate at which carbon monoxide is generated, its level in the atmosphere remains at about one part per million, a total of about 1 billion tons or four years accumulation. Dr. Lederberg has pointed out that we do not know what removes carbon monoxide from the atmosphere, thus maintaining the stable low level of contamination. As long as we do not know who or what our secret benefactor is, there is the constant danger that we will destroy it by accident. The more of our surroundings that we destroy (even unwittingly), the greater the chance that we shall destroy something vital such as the controlling factor for carbon monoxide.

The final entry in our inventory of automotive pollutants is a class of substances known as oxides of nitrogen, chemical compounds formed of nitrogen and oxygen. These compounds are in large measure unstable and, hence, physiologically active. They are irritants when breathed and when absorbed on moist surfaces. Some can cause the destruction of lung tissue. Widespread spotting of plants, especially of tobacco leaves, in Connecticut was traced to these oxides and to the vehicular traffic on the Merritt Parkway and other arterial highways.

*The Washington Post, May 24, 1969.

The number of automobiles in the United States is so tremendous that any pollutant that is thrown off with regularity can no longer be ignored — be it exhaust fumes, tires and brake linings, kleenex and cigarette butts, or the rusty, discarded body of the car itself. We now burn 80 billion gallons of gasoline, nearly all of it containing poisonous heavy-metal additives. By 1975 the gasoline consumption is expected to be twice today's level. Exhaust contaminants enter the atmosphere in billion-ton lots; the atmosphere, however, consists of only 1 million billion tons of air. Therefore, we are polluting the air annually at one-part-per-million concentrations, which are frequently high enough to have detectably adverse physiological effects.

Two facts are generally unappreciated by the man-on-the-street. First, to get rid of our atmospheric pollutants, we rely illogically on chemical or physical breakdown or on fallout. The former method, however, is not reliable; the latter, especially in the case of heavy metals or persistent organic poisons is neither effective nor desirable. Second, to an extent we are unwilling to admit even subconsciously, we Americans rely on the inability of other peoples of the world to match our industrial achievements. If the rate of pollution of the world's atmosphere were to be increased 20-fold (since the United States contains only 1/20th of the world's population) and then doubled (because our consumption of gasoline is expected to double by 1975), worldwide levels of pollution would be intolerable. This is the message that must be brought to the attention of all persons over and over and over again so that we in the United States can start rational corrective steps at once.

2. Pesticides

The amount of available water limits the quantity and, for human beings, the quality of life on earth. A plant that requires certain amounts of water at specified times will not be found in geographical areas where these needs are not met. The irrigation of a previously near-waterless desert permits the survivial of imported plants that would otherwise die. A standard of living that demands a certain amount of water per person will be attained only by a population whose numbers are small enough to allow the total demand for water to fall within the amount that is available.

The total water supply of the United States also serves as a basis for deciding if the nation's chemical industries are producing poisonous substances in hazardous amounts. A decision reached in this way does not have specific application in the sense that the life of Joseph Newcombe of Waco, Texas, is in danger; rather, it is a decision that applies to a large group of persons. To whatever extent some of these persons have escaped exposure to dangerous chemicals, others have been exposed to an extent that is correspondingly greater. The area of the continental United States is approximately 3 million square miles. The average rainfall for the United States is about 30 inches per year. Using the land area and the amount of rainfall, we can compute the total volume of water available annually to the United States as roughly 200 trillion (2×10^{14}) cubic feet. A cubic foot of water weighs 62.4 pounds; so, by neglecting some miscellaneous digits that serve only to complicate the arithmetic, we can say that the total weight of water that falls on the United States is approximately 10 quadrillion (10^{16}) pounds.

During the summer of 1969 the inhabitants of the Rhine Valley and the Dutch at Rotterdam on the Rhine River delta had an unpleasant fright. The spillage of what was reported to be 200 pounds of endosulfan (an insecticide sold under the trade name Thiodan) into the Rhine River at Bingen, Germany, caused the death of more than 100 tons of fish and many waterfowl as the poison flowed toward Rotterdam, 250 miles downstream. The city of Rotterdam switched to emergency water supplies for several days while the poisoned water flowed by and out to sea.

Endosulfan is lethal to fish at concentration of 0.1 parts per million (ppm). Water taken from the Rhine River during the week of the catastrophe was able to kill healthy fish in a matter of minutes. Endosulfan, according to reports from France and contrary to some accounts, is also toxic to human beings. During an antimosquito campaign in southwestern France, 75 cases of poisoning are known to have occurred, of which four were fatal.

For purposes of discussion, the level of endosulfan that is lethal for fish, 0.1 ppm, can be used as a maximum permissible average level of pesticide contamination, calculated on the basis of all the water of the United States. Ten quadrillion pounds of rain or snow fall on the United States each year. I now suggest that the annual manufacture of widely disseminated poisons in quantities so great that they could contaminate all of this water to a level greater than 0.1 ppm (which, incidentally, is higher than the DDT contamination level now permitted in cows' milk by federal regulation) is a dangerous practice and one that calls for serious thought and possible remedial measures.

The quantity of pesticides needed to contaminate 1 quadrillion pounds of water to a level of 0.1 ppm equals $10^{16} \times 10^{-7}$ or 10^9 pounds, 1 billion pounds of pesticides. By coincidence, a headline in the December, 1968, issue of *Agricultural Chemicals* reads: "U. S. Pesticide Production Again Tops Billion Pound Level." This headline in my estimation bodes serious trouble. Many authorities, of course, will dispute this claim by citing the irregularities of distribution of insecticides and other pesticides (including their exportation abroad), by pointing out that many pesticides are retained in the soil, and by disputing the limit of tolerance (0.1 ppm) that I have used as my yardstick.

In reply, I would claim that these persons are avoiding rather than confronting the problem. They are temporizing in the hope that the problem will go away. Irregularities in distribution merely mean that some segments of the population escape at the expense of others. The accumulation of long-lived substances in the soil does not make them less dangerous; it merely postpones the day of reckoning. Herbicides have been known to kill crops planted in fields last treated four years previously. The rate at which some herbicides are degraded by soil organisms is less than 3 per cent per year; these poisons will accumulate in the soil to levels 30 or 40 times their annual application levels. Finally, the level of 0.1 ppm that I have used as my yardstick is the very basic level of contamination; it is well known that many insecticides are concentrated manyfold as they move through the food web.

In the tidal marshlands of Long Island, primarily because of mosquito control measures of the past twenty years, DDT has accumulated to 0.04 ppm in plankton, 0.16 ppm in shrimp, 1 or 2 ppm in minnows, and about 75 ppm in gulls. Carnivorous birds have levels of DDT 10 to 100 times greater than the fish upon which they feed and about 1 million times higher than that of water itself. A striking and now well-documented effect of DDT in birds is the suppression of eggshell formation. Populations of herons, grebes, gulls, and birds of prey throughout the world have been seriously affected by the reduction, and at times complete elimination, of surviving nestlings. The brown pelican appears to have lost the ability to reproduce. The general population of the United States has about 10 or 15 ppm in its tissues; mother's milk, as many persons are now aware, contains much more DDT than the amount permitted in cow's milk under existing federal regulations.

Pesticide pollution is an especially dangerous threat to the health of the nation because those ultimately responsible, the large chemical manufactures, use the ever-increasing need for food as a justification for the continued — nay, expanded — production of agricultural poisons. Opposing food production is as unpopular as opposing motherhood; indeed, the two are not unrelated. Here, then, is a colossal example of the "tragedy of the commons." If the control of population growth is removed as a legitimate topic for discussion, we are left with the following nonsensical sequence: the population increases year by year; thus, the greater use of pesticides is needed to assure the increased level of food production.

The last statement in the preceding sequence need not be true. An ever-increasing use of insecticides has been needed largely to insure that the food sent to market show no sign whatsoever of insect or other damage; a demand for perfection in *appearance* has little to do with *quantity* of available food. Year by year, because of the origin of insecticide-resistance among target insect species, an increasing application of insecticides has been called for merely to retain a constant effectiveness. Inevitably, under these circumstances there will come a time when target organisms and nontarget organisms, including man, are hit with equal effectiveness. The chemical industry does not relinquish its influence in agricultural practices graciously; to do so would be expensive. Elsewhere I have used the term "glib assurance" in describing reassuring statements made by those who have neither the knowledge to support their statements nor the responsibility for righting the harm caused by erroneous ones. The chemical industry, through its spokesmen, is guilty of much worse than glib assurances; these people are dispensing aggressive assurances when, as the primary producers, they possess both knowledge and responsibility. They support a biased position by specious arguments; they attack their critics by ridicule and innuendo. In trade-journal editorials governmental agencies, for example, are accused of "toadying" to antipesticide groups. The latter have presented their cases; in some instances they have resorted to the courts. The governmental agencies are responsible for the welfare of citizens — you, me, and agricultural chemists — and have responded to the antipesticide arguments. "Toadying" is a sorry term to use in describing the activities of agencies responsible for the nation's health.

The accumulation of insecticides in man's body can cause unexpected secondary interactions with other physiologically active substances. Interactions of this sort are well known; the alcohol in a single glass of beer can be fatal to someone who has been exposed for several hours — for example, when cleaning a rug in a poorly ventilated room — to the common household cleanser, carbon tetrachloride. Because DDT and other poisons modify normal enzyme activity, they alter the body's response to drugs used routinely by doctors and dentists. An ordinary shot of procaine can cause a toxic reaction in a patient with certain malfunctioning enzyme systems. The industry claims that this is a matter for the individual doctor or dentist to handle, that there is no need to penalize the farmer or the general public. On the contrary, I claim that neither the widespread use of pesticides nor their accumulation in man's body is necessary and, furthermore, that neither the dentist nor the family doctor is responsible for performing assays for DDT before administering a routine local anesthetic. In addition, I claim that I am part of the general public. The quantities of poisons dispensed annually by industry have approached a dangerous level that calls for continuous public scrutiny and, whenever necessary, for the full use of governmental regulatory powers. The use of these poisons at perpetually dangerous levels calls, too, for the reintroduction of population growth as a topic for discussion. The omission of population growth from the discussion of agricultural poisons renders decision making impossible and reduces the entire discussion to an exercise in futility.

3. Thermal pollution

The auditorium in which 200 or more scientists were meeting was not air-conditioned. The heat was stifling; the room, unbearable. Suddenly the door opened and three workmen appeared carrying a window-type airconditioner. Quickly they placed the machine, cooling coils pointing toward the audience, on four chairs near the speaker's platform. The foreman plugged the machine into a nearby wall socket. The switch was thrown. Satisfied that a stream of cold air was now cooling the auditorium, the workmen left, acknowledging with cheerful nods the apparently speechless gratitude of the visiting scientists. No sooner had they departed, however, when someone arose and, to the prolonged applause of the audience, turned off the noisy, useless machine.

A window cooler works only because it deposits cool air on one side of an insulated barrier and hot air on the other; persons within an air-conditioned building can, under these circumstances, enjoy pleasant temperatures on an otherwise sweltering day. A window airconditioner placed entirely within a room blows as much heated air from one end as it does cooled air from the other; it blows more heated air, in fact, because the motor and the compressor generate heat. And so a window airconditioner placed within a stifling auditorium will only add more heat and make the room even more unbearable.

Small cooling appliances can be placed entirely within large rooms. A refrigerator tends to warm the kitchen (and the rest of the house as well) with the heat that is removed from inside the cabinet. Fortunately, the cooled volume is small and the house is large and so the heating effect of the refrigerator is scarcely noticed. Life would be unbearable for the family in one-half of a duplex apartment if the other family installed a window-type airconditioner in the dividing partition. If the two apartments were of identical size, lowering the temperature of one from 85 F to 70 F would increase the temperature of the other from 85 F to 100 F or more. Cooling devices rely upon efficient heat removal systems – either a relatively large volume of open space or, for water-cooled devices, a plentiful supply of cool water that can be heated and discharged, thus removing the heat from the premises.

Mankind, because of his numbers and his technological appetite, no longer lives in an infinite world. The world no longer provides a "relatively large volume of open space" for mankind's collective heat-absorbing needs. The story is a repetitive one but it can stand telling once more. Given man's bulk and man's needs, the world is no longer infinite. It is finite, or worse; because man and his needs increase exponentially, the world is effectively shrinking at an alarming rate. Such is the story of pollution of the environment; such too, is the story of thermal pollution.

The amount of electrical energy used within the United States has doubled every 9 years for the past 40 years. By 1980 the production of electricity in this country will be 2 million megawatts (2 billion kilowatts); its production will require for cooling purposes alone 200 billion gallons of water per day. The total runoff in the United States is about 1,300 billion gallons per day, so nearly one-sixth of the daily runoff water will be used for cooling electrical generators. The increase in the temperature of water used for cooling generators is about 20 degrees; one-sixth of the runoff water in the United States at some moment each day will be 20 degrees hotter than it was at the day's start.

On the drawing boards at the U.S. Atomic Energy Commission and in the offices of public utility firms throughout the nation are plans for 100 nuclear generating stations; these are less efficient than those that burn gas, oil, or coal and so the amount of water needed for cooling will be correspondingly larger. The amount of water required by the 100 proposed nuclear generators is reputed to be 700 billion gallons daily. These generators, already well into the planning stage, together with the old-fashioned fossil-fuel generators, would require from one-half to two-thirds of the total runoff of the United States for cooling purposes. The temperature of this daily discharge would be raised 20 degrees. A good deal of this heat would be dissipated by evaporation; nevertheless, river life and estuary life will be unavoidably altered because few living forms are able to survive and reproduce in lukewarm water. Lukewarm? One river in the Northeast has been known to reach 140 F; this temperature is near the limit that can be tolerated when tested by hand. Such would be the temperature of the water in rivers from which city after city draws its drinking water and into which each, in turn, dumps its sewage. Even viewed from today's rather soiled vantage point, the prospect seems dismal indeed.

The thermal pollution of the nation's environment comes from a multitude of sources other than electrical generators. The Atomic Energy Commission's Hanford plant alone raises the temperature of the Columbia River, a mighty river that accounts for nearly one-third of the entire river water of the United States, 3 to 6 degrees depending upon the season of year. Many industries use convenient river water for cooling purposes, pumping it into the plant at one temperature and discharging it downstream at a slightly higher one.

The enormous number of air-conditioned buildings in large cities represent a tremendous outpouring of heat. The peak demand for electrical power in the United States is no longer in the winter as it once was; the peak is in the summer when the demand is for cooling. The volume of air inside large office buildings and apartment houses is sufficiently great relative to the open spaces of streets and sidewalks that a good deal of the heat of a city's streets is heat in the form of garbage; that is, each air-conditioned building throws its heat into the street or dumps it into the nearest body of water. The consequences are scarcely any more pleasant than they would be if the floor sweepings of the trash baskets were thrown out in the same manner. Heat is a pollutant just as surely as are material substances. We are very nearly at the point where the nation's urban centers are behaving as irrationally and as thoughtlessly as the crew of workmen who set up the window airconditioner entirely within the crowded auditorium. The outside world is becoming too small to cope safely and comfortably with the heat we attempt to throw away. The rivers are too small to handle that heat which will be generated by power plants and industry; the streets are too narrow to handle that which is pumped out of apartments and offices by window conditioners. Airplanes flying at several thousand feet bump perceptibly as they fly over large cities. The world has come to resemble the auditorium described in the opening paragraph of this essay. The heat that is produced by the multitude of cooling operations is not being carried away; on the contrary, it is merely being blown in another direction within the same room. Today he would not yet be appreciated, but on some tomorrow there will likely be prolonged applause from an appreciative audience for the man who turns off this increasingly dangerous, increasingly useless heat-making machinery.

4. Antibiotics

In 1955 a Japanese traveler returned home from Hong Kong ill with dysentery. He carried the first multiple-drug-resistant strain of shigella (the organism that causes dysentery) to be recorded in Japan. This strain of shigella was resistant to four drugs: the sulfonamides (the "sulfa" drugs), streptomycin, chloramphenicol, and the tetracyclines. By 1959 about one person in ten suffering from dysentery was infected with a multiple resistant strain of shigella; in 1964 well over half of all isolates tested proved to be resistant to the four drugs.

The human intestine and colon are "home" for a large number of organisms. A number of these are benign, if not actually useful from man's point of view. Newborn babies quickly acquire their intestinal flora from contamination during nursing and handling; should this not be so, live *Lactobacillus acidophilus*, the organism used in the preparation of buttermilk, can be purchased at the drugstore and added to the baby's formula. Changes in the composition of the natural intestinal flora — most often an increase in yeast and fungi at the expense of bacteria — following an extensive antibiotic treatment is frequently accompanied by an actue discomfort that vanishes only as the normal balance of nature within the gut is restored.

Escherichia coli, the common colon bacillus, is an ubiquitous inhabitant of colons throughout much of the animal world. Because of its almost complete lack of pathogenicity, *E. coli* is also the favorite experimental material of microbial geneticists and molecular biologists throughout the world; most of the pioneering work on the genetics of drug resistance was done with *E. coli*. The information gained from these bacteria has served repeatedly in guiding subsequent research on more pathogenic and, hence, more dangerous organisms.

During the years from 1955 to 1964 when multiple-drug-resistant strains of shigella were increasing rapidly in Japan, the frequency of multiple-drug-resistant strains of *E. coli* increased in precisely the same manner. These strains of *E. coli* were isolated not from persons ill with dysentery but from normal healthy persons chosen at random from the Japanese population. Strains isolated from persons suffering from chronic urinary infections were especially likely to be multiply resistant; infection of the bladder is one of the few pathogenic infections caused by the colon bacillus, *E. coli*. The purpose of this essay is to reveal the extent to which antibiotics have entered man's environment and the consequences of their wholesale distribution; to understand these consequences, however, we must understand a bit of bacterial genetics, especially that relevant to drug resistance.

The effectiveness of modern antibacterial drugs lies in their ability to stall the machinery by which bacteria grow and reproduce. The machinery of life, although complex in detail, is rather simple in outline. Vital processes consist of a network of interlocking and interacting chemical reactions mediated by catalysts called enzymes and designed to release energy from nutrient substances, to remove toxic waste products, and in the meantime to make and assemble more of the machinery that is capable of mediating vital reactions. Enzymes, large protein molecules consisting of precise linear sequences of hundreds of subunits (amino acids), are synthesized according to directions carried in threadlike molecules of DNA (deoxyribonucleic acid), self-replicating genetic material. Utilization of the directions carried by DNA calls for recopying them onto an intermediate nucleic acid (messenger RNA) and then translating them into the proper sequence of amino acids; the latter process involves the attachment of each molecule of messenger RNA to intracellular structures called ribosomes.

Different antibiotics and other miracle drugs interfere at specific and strategic places within the process outlined above. Erythromycin and actinomycin inhibit the copying of directions from DNA to RNA, including messenger RNA. Streptomycin prevents the proper attachment of messenger RNA to ribosomes. Puromycin is mistaken for an amino acid and is inserted into enzyme molecules, which then drop from the ribosomes as incomplete protein molecules. Penicillin interferes with the synthesis of new cell walls and in this way prevents cell division.

Drug resistance on the part of bacteria consists of countermeasures that thwart the action of the antibiotic or other drug. Changes in cell walls, for example, may suffice to keep the drug from entering the cell. An enzyme that makes a minor alteration in puromycin converts it from a dangerous drug to a welcome nutrient for the bacterium. A large proportion of the alterations that lead to streptomycin-resistance also makes the bacterial cells thoroughly *dependent* upon the presence of streptomycin for proper growth. Penicillinase is an enzyme synthesized by pencillin-resistant bacteria; it degrades penicillin into a harmless compound.

Genetic traits ordinarily are handed down from parent to offspring; in the case of dividing bacteria, a genetic trait is transmitted through the line of descent starting at its point of origin (by mutation) to all descendant cells. Bacteria, however, have other means of transmitting genetic information from individual to individual. The DNA released by a disintegrating cell may enter a second cell and become incorporated into its genetic apparatus. Viruses (bacteriophage) that reproduce within bacteria can pick up and transport fragments of the bacterial DNA from one cell to another. Cells, upon collision, can adhere and exchange genetic material in a true sexual process. Finally, small fragments of DNA can exist free of the regular chromosome in the cell but nevertheless contribute to the cell's enzymatic machinery. These fragments are known as infective particles because, like chromosomes themselves, they can pass from one cell to another.

The multiple-drug-resistance of *E. coli* is carried almost exclusively by infectious DNA particles. Precisely how the genes are assembled to form these particles is not understood; no genes other than those necessary for drug resistance seem to be included in the assembled particles. The rate at which the infectious particle is reproduced within the bacterium exceeds the generation time of the bacterium itself; consequently, a wave of drug resistance can pass through a bacterial culture faster than bacteria reproduce. The infectious particles pass from cells that possess them to those that do not; the latter then acquire drug resistance. Even though the number of cells carrying infectious particles may be low initially, the number of particles available for sharing quickly equals or exceeds the number of bacteria because the infectious particles have the shorter generation time.

Although *E. coli* appears to be the species best capable of assembling infectious particles, the spread of these particles is not restricted to *E. coli* alone. Infectious particles can pass from *E. coli* to shigella (dysentery), salmonella (typhoid and paratyphoid fever, gastroenteritis), klebsiella (pneumonia), serratia, proteus (diarrhea), and other colon bacteria including those responsible for cholera and other diseases, such as the plague. Here is the explanation for the parallel increase in the frequency of multiple-drug-resistant strains of shigella and *E. coli* in Japan from 1955 to 1964; the two species of bacteria simply shared a common pool of infectious particles.

Drug resistance in *E. coli* is of some importance to man. In the case of bladder infections that are caused by *E. coli*, the resistance pattern of the bacteria involved determines the response, or lack of it, of the infection to treatment. In the case of intestinal flora, the resistance of *E. coli* determines in large measure the stability of the floral composition to antibiotic treatment; a mixture of sensitive *E. coli* and resistant shigella, for example, would lead to an extremely severe diarrhea following the use of antibiotics because, with the removal of *E. coli*, the sole bacterial survivor would be shigella, the organism responsible for dysentery.

The presence of multiple-drug-resistant particles in the other organisms listed above poses a tremendous public health problem. These organisms cause serious, sometimes fatal, diseases; antibiotics and other miracle drugs capable of arresting them are invaluable components of man's medical arsenal. Any act that tends to multiply the frequency and types of infectious particles responsible for conferring on these organisms simultaneous resistance to large numbers of antibiotics is a disservice to the community of man.

Those of us who have had penicillin shots for streptococcus infections ("strep" throat) or for other illnesses tend to think of antibiotics as an adjunct to the practice of medicine; our view of antibiotics is restricted by our visits to the doctor or to the local drug store. Nevertheless, in the United States antibiotics are used for more than the treatment of human illness. If we scan agricultural trade journals, we can discover how far these drugs have surreptitiously entered our lives. We learn, for example, that animal nutritionists recommend *daily* doses of antibiotics as follows: (1) for pregnant ewes, 60 mg per head for 80 days; (2) for calves, 50 mg per head, to be increased to 300 mg for 15 days at change of feeding, later reduced to 60 mg; (3) for pigs, 250 mg per head; and (4) for poultry, 1 to 3 mg per bird, or 2 g of streptomycin for every gallon of drinking water. These are doses recommended by animal nutritionists.

These dose rates must be considered in conjunction with the numbers of animals involved. The United States produces 2½ billion broilers per year. There are nearly 100 million beef cattle and some 15 million dairy cattle in the country. Hogs and sheep together number about 75 million. There are over 100 million turkeys. Combining these numbers with the recommended daily doses suggests that the total annual use of antibiotics for agricultural purposes either amounts to, or would amount to if recommendations were followed, 10 million pounds or more. This same figure can be arrived at by dividing $200,000,000 (the amount of money spent for antibiotics by farmers) by 4 cents (the price of one gram of agricultural-quality antibiotic). Apparently, 10 million pounds of antibiotics *are really being used* annually for agricultural purposes.

Combinations of antibiotics are frequently used in supplementing an animal's feed. The intestinal bacteria of farm animals (including poultry) are subjected to tremendous pressures to develop resistance to these drugs. Among these bacteria is the ubiquitous colon bacillus, *E. coli*, the species that seems to be most efficient in assembling multiple-drug-resistant infectious particles. From

E. coli the infectious particles can enter salmonella (between one-half and one-quarter of all chickens at the supermarket are contaminated with salmonella), shigella, or some other colon bacterium. Since the colon bacteria of different animal hosts are constantly being interchanged, a particle-carrying *E. coli* developed within a chicken can enter a human intestine and there transmit its drug resistance to those bacteria more commonly restricted to man. The massive numbers of farm animals that are continually receiving antibiotic treatment as a dietary supplement and the tremendous amounts of antibiotics consumed by farm animals make it likely by chance alone that infectious particles conferring drug resistance, in the case of colon bacteria, will arise in the intestines of farm animals rather than in those of human beings. The actions of farmers and feed salesmen promise to negate any precautionary measure that the medical doctors might initiate in the antibiotic treatment of human patients.

The massive use of antibiotics in the feeding of farm animals constitutes a pollution of man's environment; furthermore, it threatens his health by encouraging the origin of multiple-drug-resistant strains of many different disease organisms. The original pollution, in short, gives rise to self-amplifying dangers! The public health problem arising from this pollution is an extremely serious one, and so one might expect to find that the underlying explanation for the agricultural use of antibiotics is a correspondingly compelling one. On the contrary, the excuse for supplementing the diet of farm animals seems to be both specious and trivial. Most published articles stress the financial gain in terms of added weight of marketable animals over the cost of the antibiotics; in one instance, for example, a gain of $5 to $6 was predicted for every dollar spent on antibiotics. Here is the "tragedy of the commons" presented in diagrammatically simple form. The fivefold or sixfold return on dollars spent for antibiotics will accrue only to the first few farmers who seize the opportunity to use the dietary supplements. When all or nearly all producers are using the same technique, the profit vanishes because there now arises a general overproduction that lowers the market price. Nevertheless, no one at that late moment can afford to revert to earlier, less efficient procedures. The farmers, except for a few innovators, do not gain. The population as a whole loses. The persistent winners are the drug manufacturers. It is in response to situations like these that the government, as the spokesman for the general population, should (and sometimes does) assert its regulatory powers.

5. Industrial pollution

Ordinarily, as we look about us, we do not realize how limited the natural resources at our disposal really are. In most parts of the United States water seems to be in plentiful supply; the failure of the water supply to any city in the country is so rare as to be front-page news. Air? How can air be in short supply as long as we can look up and see a sky as large as all space?

These intuitive feelings are extremely misleading. The touch-and-go use of water in the United States has been a recurrent theme in these essays. Daily needs are met over vast areas of the country by tapping groundwater too assiduously. Water levels have dropped. Many streams that used to run throughout the year today run only as muddy torrents immediately after heavy rainstorms. The water table has dropped to levels too low to maintain as permanent waterways streams that once compared in size to what are rivers in many other lands. The permanent streams that still remain are tapped repeatedly by cities for drinking water and by industry for cooling purposes; they are polluted repeatedly by city sewage, by the private trash of thoughtless individuals, and by industrial wastes of all sorts.

At frequent intervals the supply of air, to the amazement of some, to the sorrow of others, and to the apparent indifference of most, is also in surprisingly short supply. The reason is the formation of layers of air that prevent the dissipation of smoke and other noxious wastes. These *thermal inversions* form daily in some areas of the country, at more irregular intervals in others. To argue under these circumstances that the smoke from a chimney has a world of atmosphere in which to dissipate is a grievous error; the smoke collects within a stationary bubble of limited diameter and of limited height. The danger of pollution, therefore, is related to the rate at which pollutants are fed into the atmosphere and the *smallest* bubble that is encountered at rare but nevertheless predictable intervals. The smallest bubble is the controlling one because, as Mark Twain once suggested, all the problems of the world could be solved by turning off the oxygen for five minutes.

Where in the United States does a layered structure of the atmosphere occur? Where are these thermal inversions that govern the severity of air pollution? Everywhere! The number of hours of every hundred spent under an inversion varies from 10 along some portions of the Atlantic and Gulf coasts to 50 in some regions of the Rocky Mountains. During the winter in the Great Basin, an area that includes all of Nevada and portions of its bordering states, inversions may persist for as long as 10 days at a time. Cities tend to be built where inversions are especially common. Los Angeles is under an inversion 270 days of every year. The exhaust fumes of downtown Los Angeles traffic, 4 million motor vehicles, can at times rise *no higher than 30 feet.* And yet in safety campaigns we emphasize over and over that an automobile should never be started in a closed garage! The Hudson, Monongahela, Allegheny, and upper Ohio Rivers – all heavily industrialized – are enclosed by frequent and persistent atmospheric inversions. I include herewith accounts of disasters caused by industrial air pollution at Donora, Pennsylvania, and in London, along with descriptions of pollution in the Raritan River in New Jersey and the Rhine River.

On October 30, 1948, a smog that had settled over the Monongahela River Valley at Donora, Pennsylvania, four days earlier became particularly irritating and nauseous; before the day ended, 17 persons had died and nearly 6,000 others (in a city of less than 14,000) had become ill. Three additional deaths occurred during the following few days.

An investigator who arrived in Donora during the next week found a zone of almost complete destruction of all plant life beginning at a zinc smelter at the edge of town and extending several miles eastward into the neighboring hills. This was a zone of desolation several decades in the making; its origin could be traced to the construction of the zinc smelter in 1916. All trees and bushes in the city's cemetery were dead; the ground was so eroded by years of uncontrolled runoff that dogs would occasionally retrieve human bones. What was originally intended to be a city park also lay in the path of the fumes and was instead a barren wasteland.

During the late October smog, many farm animals died in the rural areas surrounding Donora; a census of these includes 14 sheep, 2 pigs, 740 chickens, 12 colts, and 6 cows. More livestock would have been killed during these days except that farmers long ago had learned that the fumes were destructive under the best of circumstances and that farming in the areas immediately surrounding Donora was simply impossible.

The events responsible for the death and destruction at Donora were a combination of the industrial exploitation of an environment to its utmost and a rare inversion, more severe than most, that made "utmost" a grossly inadequate term. Conditions that had been scarcely tolerable at best became lethal under the more restrictive atmospheric inversion of late October. Donora lies on the Monongahela River between 300- and 400-foot bluffs. At the river's edge were four sets of railroad tracks, the zinc smelter, a steel plant, and a wire manufacturing plant. Each plant perpetually belched out red, black, and yellow smoke. The factories used dozens of small coal-burning switch engines to manipulate the flow of material into, through, and out of the various plants. The zinc smelter also produced sulfuric acid and, largely because of inefficient manufacturing processes, large quantities of the acid were expelled through exhaust stacks. These were the industrial activities that kept Donora in a perpetual haze of smoke and fumes; this was the level of air pollution, known to be fatal to plants and farm animals, tolerated as a way of life by man. When the atmospheric inversion literally capped the river valley so that the fumes could not escape, the contaminating poisons quickly reached concentrations that were lethal to man as well.

Events comparable to those that struck Donora in 1948 have hit London periodically for decades or longer. Again, the trouble arises when the volume of air available to the city becomes restricted by stagnation under an atmospheric inversion. All of Greater London's smoke (much of it sulfurous soft coal smoke as in John Evelyn's London of 1661) then becomes trapped in the large shallow bowl of the Thames River within which the city sits.

In 1948, one month after the Donora disaster, smog settled over London for five days, during which the number of deaths was 450 above normal, these extra deaths can be ascribed to the smog. In December, 1952 the scenario was reenacted; four days of extremely heavy smog claimed 4,000 lives. In 1956 a third period of smog claimed 1,000 victims, and in 1962 a fourth took 750. The decrease in the number of deaths in the 1956 and 1962 smogs was the result of public health measures that were put into effect when disaster threatened: invalids and elderly persons were urged to stay indoors, to rest as much as possible, and to avoid physical stress; persons going out were urged to wear breathing masks; and all activities, industrial or individual, that tended to pollute the air were halted as the level of contamination rose. Today, conditions in London have improved; oil has been substituted for coal. So, in 1970, November passed with no smog for the first time in centuries.

The basic problem remains largely untouched however: the pollutants — visible or invisible — that are discharged into the air by a heavily industrialized society are too great to be swept away by air currents. Bubbles of air are formed by inversions. These bubbles have boundaries that are as real as if they were contained within modernistic geodesic domes (such as Buckminster Fuller proposes to construct over Manhattan!) and are too small to absorb the outpourings of modern industry. Pained resentment arises when those in charge of industrial plants suggest, in effect, that those persons living nearby occasionally hold their breath for a few hours, or days; the dangerously restrictive inversions, according to many industrial managers, are acts of God whose damaging effects are not to be blamed on industry.

I could recite more accounts of air pollution, such as the story of beryllium particles causing lung disease in persons living several miles from their source in a fluorescent lamp factory or how the residents of Birmingham, Alabama, breathe as many cancer-forming substances in a single day as one would obtain by smoking 2½ packs of cigarettes. I must hasten, however, to another aspect of industrial pollution, that of our rivers and waterways.

The Raritan River of northern New Jersey can be cited as a river that has been converted from a sportsman's retreat into a discolored stream of industrial wastes. According to a report issued from the White House in 1965, the waters of the Raritan River contain phenols at concentrations of nearly 1100 ppm (parts per million); formaldehyde, nearly 2000 ppm; copper, 6 ppm; and arsenic, 375 ppm. The same report explains that 3 ppm of copper is fatal to some fish within 48 hours; pennies that are tossed into public fountains containing goldfish will quickly kill them. Phenols (Lysol is a phenolic antiseptic), formaldehyde, and arsenic are extremely toxic compounds. The Raritan has become a desolate river that offers little pleasure to people living nearby and no chance of survival to fish or other forms of life that might otherwise inhabit it.

Raritan Bay, once a resort center noted for its swimming and fishing facilities, has been rendered useless by the pollution from the Raritan River watershed and from other streams that flow into the bay area. In addition to the heavy concentration of organic sewage brought by all streams (including the Raritan River itself), the Arthur Kill brings in lethal pollutants from oil refineries as well.

The Rhine, one of Europe's largest and most beautiful rivers, has become a giant sewer flowing from Switzerland, along the border of France, and through the industrial heart of West Germany to Rotterdam and the North Sea. Accounts of the pollution of the Rhine were widely publicized during the summer of 1969 when it was accidentally contaminated by an insecticide, reportedly Thiodan; an estimated 40 million fish were killed in this one accident.

The industrial pollution of the Rhine has a thousand sources. On any one day 10,000 riverboats operate on this river plying between ports of Germany and France and Rotterdam, the world's busiest seaport. Each of these boats carries the captain's family as well as its crew; the population of river folk on these boats alone must amount to nearly 100,000 persons.

Near Strasbourg, France, potash plants dump some 30,000 tons of waste salt into the Rhine each day. Here more or less begins the lengthy pollution of the river. As it flows through Karlsruhe, Mannheim, Mainz, Koblenz, Bonn, Cologne, and Duisburg, the Rhine flows through a valley comparable in industrial capacity to that of the upper Ohio River (and its branches) in the neighborhood of Pittsburgh and Wheeling. The industrial region along the Rhine is, if anything, more compact than the corresponding region of the Ohio River. The wastes of German industry together with the groundwater that is pumped from the mines of the Ruhr Valley into the Ruhr River are poured into the Rhine River for convenient disposal. In a recent test, a small white card proved to be visible through nearly four feet of water at the Swiss border where the Rhine first enters Germany; at Duisburg the card could be seen through a bare six inches of water. Such is the effect of dumping the wastes of 30 million persons into one river — even when the river is one of the world's largest.

6. Concluding remarks

Those who pollute the environment seem to adopt one or the other of two defensive stances: they claim either that they are in fact not causing pollution or that the pollution they are causing is but a small price to pay for industrial progress. In an angry account of the Donora disaster, Dr. Clarence A. Mills reports:*

*Clarence A. Mills, *Air Polution and Community Health* (Boston: The Christopher Publishing House, 1954), p. 44.

On Sunday, October 31, 1948, the afternoon's gentle rain washed the deadly poisons from Donora's air and brought relief to the stricken thousands. Then began a second smoke screen, one of claim and counterclaim, of blame shifting to hide the tragedy's ugly and sinister significance. Even with their long history of atrocious pollution of the valley air, zinc and steel plant officials now made vigorous denial that output from either of their plants could have caused the disaster.

One symposium after another, according to Dr. Mills, was used to sound the refrain, "There is no proof, there is no proof."

Attitudes and practices have not changed a great deal in the intervening years. The tobacco industry and their supporters respond vigorously each time the Surgeon General issues a report linking smoking to lung cancer, heart disease, and emphysema; in each case we hear, as did Dr. Mills in Donora two decades ago, "There is no proof, there is no proof." The *New York Times* reported recently, "TV Makers Act To End 'Excitement' Over Radiation." "No hazard situation exists and there is no justifiable cause for public alarm," one official reported. Nevertheless, until the color television sets were corrected, large segments of the population were being exposed needlessly to a low level of nearly continuous X-radiation. Spokesmen for the chemical industry are heard to comment that they are tired of hearing of phosphate in lakes and rivers. The phosphate in question is obtained at great expense (in terms of energy required) from ores, is transferred by the magic of chemistry into household detergents, and almost immediately is swept irretrievably into lakes and rivers where it supports an excessive amount of plant life; these plants when dead and decaying pollute the water and smother nearly all higher forms of animal life from crayfishes on up. Finally, at a symposium held a scant month after the Santa Barbara channel oil "spill," an expert on marine ecology told conference members that he had encountered a surprisingly low mortality of marine plant and animal life, that the incident had not been as catastrophic as early reports had indicated. In one survey of industrial executives, about two-thirds denied that air pollution is either a problem or that it affects people's health; over one-third claimed that the problem exists only in the minds of some people.

Ecology has become the "subversive science" in the eyes of industry because ecologists are the persons whose minds are most disturbed by pollution and its far-reaching effects. When the issue of thermal pollution halted the construction of an upstate New York nuclear reactor, those who initially raised the objections were labeled "so-called experts" and "do-gooders." When scientists called for a more thorough discussion of the space agency's proposal for avoiding contamination of the earth by extraterrestrial life, they were called "longhairs." When someone expresses concern over the growing accumulation of DDT in higher organisms, including man, he is told to ask the starving Indian child how much DDT *he* has accumulated in his body's fatty tissues. Rachael Carson, whose *Silent Spring* has been more effective than any other publication in alerting persons to the matter of pesticide pollution in man's environment, is periodically reviled in pesticide trade journals. Indeed, one spokesman for the chemical industry suggested in an article published in *BioScience* (1967) that those who oppose pesticides may be preoccupied with the subject of sexual potency.

The attitude of business executives toward pollution and the arguments used by their spokesmen are reminiscent of the old but wonderfully absurd expression, " 'Shut up,' he explained." The tide of public opinion, however, must be turning even if ever so slowly. We learn from the *Public Relations Journal* that in matters of pollution company officials are well advised *not* to call complainers "troublemakers" or "pinks"; nor, when the evidence is plainly against them, are they to deny the facts. Wonderful! Most encouraging of all, perhaps, is the further admonition not to take comfort in the notion that complainers who are also employees will not carry their complaints too far. Kick and scream as they will, executives will be dragged into this century!

Much of what is said about pollution's being the price of progress has a small, remarkably small to be sure, kernel of truth — if one maintains a limited view of the problem. The demand for electrical power has doubled every 9 years during the past 40 years; in order to maintain this rate of growth, thermal pollution must be accepted. The demand for automobiles and trucks grows every year, and it will grow even more when the government succeeds in its program for mass producing and scattering broadcast single-family dwellings. Air pollution by automobile exhaust fumes must, therefore, increase. Air traffic has clogged existing airports, so many new ones must be built; noise and aircraft exhaust pollution will increase of course.

Behind all of this resigned acquiescence to worsened conditions as a necessary way of life are two assumptions: (1) that the number of persons is increasing and, presumably, should continue to do so and (2) that the amount of energy placed at the disposal of each individual should be unbounded. I dispute both of these assumptions. If they are accepted, they can lead only to disaster. Indeed, as in the case of the potato famine of Ireland, the time for taking corrective measures will probably pass unnoticed — then — disaster will strike! By this I mean that the wave of industrialization will sweep through all nations of the world at a time when the total number of persons times the energy demands of each will equal a devastating drain on and pollution of what are erroneously regarded as limitless resources.

To many persons, the alternative to today's industrial and technological society is a return to the arts and crafts of Renaissance Europe. This is not the only alternative; indeed, if it were not that the present road is one to certain disaster, it would not even be an *acceptable* alternative.

Provided that population growth can be brought to a standstill in the very near future, modern industrial techniques can be tolerated or even expanded. We need, however, a population of citizens, teachers, technicians, and politicians who understand what may be called the "newer math," a mathematics that includes the notion of a finite earth in its calculations. Materials used for any process are drawn from a finite source that must not be exhausted; waste products produced by any process are discharged into a finite world that must not become overly contaminated. The "newer math" should apply with equal vigor to the problem of disposing no-deposit, no-return bottles as to the question of the total energy and effort required for interplanetary exploration − not omitting the consequences of beryllium pollution from rocket exhausts. Reasonably simple calculations may reveal that the world can ill afford either the bottles or the interplanetary exploration. So be it!

The Environment and Technology Assessment *

Milton Katz

To save the environment, America will have to make far better use of its technology. "Technology assessment" in a new form can provide a way to do so.

Technology assessment in some form has long been a part of industrial society . . . [It] occurs when a business enterprise estimates the costs and gains expected from an investment that would introduce a new technology, or expand the use of an existing technology . . .

Under present assessment practice, some important costs are overlooked. This is convenient for the particular business . . . but bad for the environment.

An electric power company, for example, in trying to decide whether to install a new power plant, will treat the production and sale of additional power as the prime objective or "benefit." It will treat the fuel burned in producing the power as a cost, but not the smoke that may pollute the surrounding air nor the waste products which may be discharged into nearby streams. The damage to the community caused by the smoke or other waste products is treated as "social cost," not a cost of the business enterprise; it is regarded as an "external," not an "internal" cost. . . .

"Technology assessment" is needed in a new form and with a new emphasis. The new "technology assessment" would alter the way in which the "costs" and "benefits" of technology are calculated. [Business enterprises would not] be permitted to ignore such "side effects" of technology as air pollution, water pollution, sonic boom, or jet noise by brushing them aside as "external." The "side effects" would be included as costs in realistic and balanced cost-benefit calculations.

The objective is to reduce harmful side-effects to a minimum while preserving the positive contributions of technology. . . .

Our immense technological endowment alone is not enough. To realize the potentialities, the American people will also have to care: care enough to do something about it. They can press their political leaders to act. Through the courts, they can invoke the help of familiar but steadily evolving doctrines of common law. Through technology assessment in the new sense, they can put their cares to work.

•   •   •

Personal Commentary

The new form of "technology assessment" advocated by Professor Katz corresponds in part at least to the "newer mathematics" of my preceding essay — a mathematics that recognizes the finiteness of the earth from which resources are drawn and into which pollutants are poured. In order to correspond even more completely with my "newer math," technology assessment in its new form must include in its list of potentially harmful (but possibly beneficial) side effects the impact of proposed technologies on the established flow of *materials* through the older technologies that are to be displaced.

Toward Solving
Pollution Problems

Pollution, because of its many causes, takes many forms. Garbage accumulates because it is — momentarily — cheaper to discard bottles, used cars, and old refrigerators than it is to return them for reuse. Manure collects on farms because artificial fertilizer is cheaper to buy and use; hauling manure to fields and spreading it requires expensive labor. Industrial wastes pour into rivers or rise into the air because they are unwanted by-products of an otherwise profitable commercial enterprise.

Wastes will not spontaneously disappear from our landscape, our rivers, and our air. They will disappear only because someone has fought to get rid of them. Hence, the emphasis on *our* landscape, *our* rivers, and *our* air. These will be cleaned up when it is worth *our* while to fight and when it is worth *our* while to forego the trivial and peripheral luxuries responsible for much of life's litter.

Learned panels meet to discuss the deterioration of man's environment. Distinguished scientists wrestle with ways of solving the pollution problem. They issue reports to Congress, to administrative agencies, to industry, and — through the press — to the people. Those reports that deal with techniques and procedures for initiating *change* make even gloomier reading than do the others that merely rail at pollution with evangelical fervor. The panels charged with implementing policy decisions realize that people do only what people want to do and, to a depressing extent, they want to do what is expedient, what pleases or profits them at the moment. No panel of 12 or 15 men, however distinguished, can stem the flow of wastes of 200 million persons.

Education of the public offers the only lasting solution to the pollution problem. Here, too, reports have been gloomy. Those who predict serious trouble for mankind, it is said, lack convincing evidence; arguments about and predictions of future events are supposititious. There is, the argument continues, no sound basis for predicting what a given technological innovation will lead to or what problems it will pose. And so once more gloom and pessimism settle in: the people will change when they have learned their lesson. Only by then it may be later than we think and the lesson, like the famine of 1840 in Ireland, may be more costly than we wish. This essay touches on the matter of predicting the outcomes of certain innovations. Means of predicting do exist, we shall see, and they do offer a basis for forestalling many types of pollution.

Living organisms provide a model for the use and reuse of material goods. Nutrients, including carbon for fuel, go into the body together with oxygen; out of the body pass nitrogenous wastes, carbon dioxide, and water. In between occurs a remarkably complex series of chemical reactions, each supplied by materials obtained from previous ones and each in turn supplying material to the next one. After a half-century's painstaking work, biochemists have unraveled nearly all of these interlocking and interconnected reactions. Immense charts illustrating them are available from industrial firms; these charts are status symbols in many laboratories where they make delightful wall decorations. They are useful decorations, too, because from the lines depicting the flow of reactions in living cells, we can foretell with fair accuracy the outcome that would follow the blockage of any single reaction. By placing his finger at one spot on one of these diagrams, an observer sees which materials will accumulate, which will be depleted, which reactions will be speeded up, and which will cease unless they are fed by subsidiary routes; indeed, the charts help identify these metabolic bypasses.

The flow of materials through the economy of a nation, I contend, is no more complex than is that of the myriad of biochemicals in a living cell. Effort and library research alone should suffice to enable one to sketch the flow of the materials of commerce. An efficient economy will show many closed loops resembling the recycling of materials planned for long-range space explorations. An inefficient economy will show numerous one-way paths ending as "waste." The terms *"efficient"* and *"inefficient"* are applied to these economies by definition despite the dollar price tag of the moment; to throw irreplaceable resources down the drain is expensive. To convert, for example, 3 million tons of phosphate ore annually into a half-billion pounds of phosphorus at the expense of nearly 3 billion kilowatt hours of electricity and then to throw this phosphorus down the drain – literally – as a phosphate detergent the first and only time that it is used is too expensive for any nation to afford. Phosphorus is not all that common; lakes and rivers are not all that large; and fossil fuels are needed for other purposes.

Two charts that illustrate the efficiency and inefficiency of one small facet of our economy – the fate of the pop bottle – are shown here to illustrate the cyclic flow of the earlier, returnable bottle and the one-way flow of the no-deposit, no-return bottle. A child can see the difference between the two charts; a child could have foretold the consequences of breaking this cycle. Momentarily, the change to the no-deposit, no-return bottle brings prosperity to the bottle maker and his employees, to the storekeeper (because he saves on labor costs), but it increases the cost of garbage removal – a cost born by the consumer and his neighbors. In the long run, however, the broken cycle means that the *persons* who work in the factories, in the stores, and in the communities will be surrounded by more garbage and trash than they need be. In the end, *persons* are the ones who suffer and so, despite momentary shifts in apparent economic gain, the no-deposit, no-return system is inefficient.

The flow of materials should be forced into closed cycles by government regulation. "No-deposit, no-return" should be treated as a four-letter word. The cost of disposal by means of closing the cycle should be included in the price of every car, range, refrigerator, or furnace manufactured. At those junctions in the flow chart of commerce where the recycling tendency is weak, subsidies should be provided to make the cycle firm. A subsidy of this sort would serve as a valve that regulates the flow of material; it would be no more expensive than the eventual cost of gathering garbage and trucking it to an overcrowded and ever-growing city dump.

Every new technological process should be examined not only for its direct effect on public health but also for its effect on the existing flow of materials. Should the existing flow be severely upset, the new process should be prohibited or forced to share in the cost of rebuilding old cycles into new ones. If artificial chemicals displace manure as a natural fertilizer, and if the need for milk, meat, and poultry products makes inevitable the production of animal wastes, then the introduction of artificial fertilizers is not an unmixed blessing. The repair of pre-existing cycles should be one of the burdens to be shared financially by the artificial fertilizer industry — to be passed on, of course, to the ultimate consumer. The point is that tremendous quantities of waste materials have been diverted from a natural cycle of formation and destruction while no satisfactory provision for their eventual disposal has been provided.

A similar story could be developed in the case of the soap industry. As late as 1949, 95 per cent or more of all soaps were made of agricultural materials; by 1965, 80 per cent or more of all "soaps" were synthetic detergents. The introduction of detergents into everyday household use has been accompanied by a series of problems. First, the early detergents were virtually indestructible, so rivers, and eventually drinking water, began to foam. This defect was corrected under laws that demanded that detergents be *biodegradable,* that they be subject to degradation by soil bacteria. Second, the new biodegradable detergents are synthesized from phosphorus compounds. At the moment, as I have previously pointed out, this represents a very wasteful and disruptive pathway in our economy; a great deal of energy is expended on a process that immediately causes a severe pollution of our lakes and rivers. Third, artificial detergents are now strengthened by the addition of proteolytic enzymes, which are especially useful in removing blood stains not from men's shirts as the advertisements claim but from ladies' underwear. Unfortunately, these same enzyme-containing detergents have been responsible for many allergic reactions among sensitized persons. These are the problems that come with the use of synthetic detergents; they are the problems that correspond to the sins of *commission.* There are other problems, those corresponding to sins of *omission,* that come with the failure to use animal fats in today's soap industry. What has become of the animal fat that once went into soap? Some is finding its way into bird feeding stations. If any of it is accumulating as unwanted waste material, then the synthetic detergent industry must be blamed for destroying what was once an operating cycle of use and reuse.

Obviously the world need not operate exclusively by means of cyclic processes. Nevertheless, cycles are the logical pathways by which materials flow within a self-contained system — intracellular metabolism, general physiology, space-capsule technology, or the maintenance of life on earth. Modern technology poses problems of various sorts. The cyclamate affair of 1969 was one of these. The use of cyclamate did not destroy any preexisting sugar cycle; it merely proved to be too dangerous a chemical for human consumption. Artificial colors that prove to be cancer-causing agents and other chemicals that are used for trivial reasons are a class in themselves. For them, society can do no more than compare the need for the chemical with the harm it may do. In my estimation, the possible error in judgment should always be on the side of caution rather than on the side of the innovator whose motivation is personal financial gain.

Since we recognize that the accumulation of wastes in the environment imposes a burden on the entire population that could be lessened if the wastes were collected and concentrated at their source, it seems that a detailed flow chart of materials of commerce would be an extremely valuable document. Governmental subsidies, as I mentioned earlier, could act as valves in controlling the flow of materials. Sulfuric acid that is produced as a waste product in one plant and is allowed to flow into a nearby river could be collected and shipped, with the financial support of public funds that would otherwise be spent on water purification, to an interested consumer. In other words, wastes of one firm could be forced by subsidy to reenter the commercial traffic of the nation as a whole. If the need for recycling of this sort and for the avoidance of unnecessary waste does not soon register with the majority of citizens, mankind will be in serious trouble. We shall be in the same sort of trouble as a crew of astronauts would be if they were to be sent on a trip to Mars carrying ham sandwiches, a water cooler, and a chamber pot. The scale may be larger in the case of a nation's economy but the ingredients — before and after consumption — are much the same. And it may become very nasty underfoot indeed.

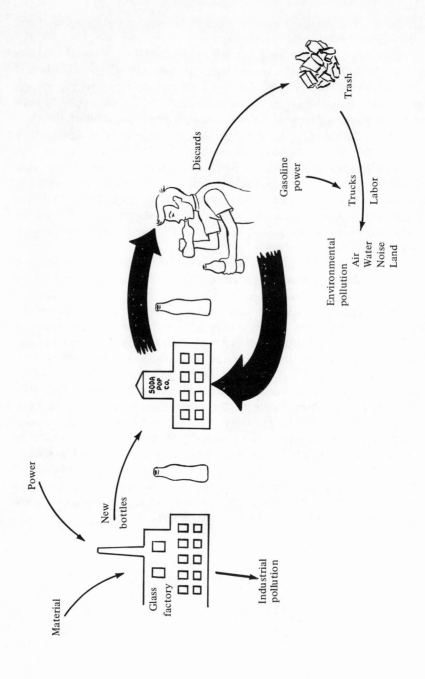

Material

Power

Glass factory

New bottles

Industrial pollution

SODA POP CO.

Discards

Trash

Gasoline power

Trucks

Labor

Environmental pollution

Air
Water
Noise
Land

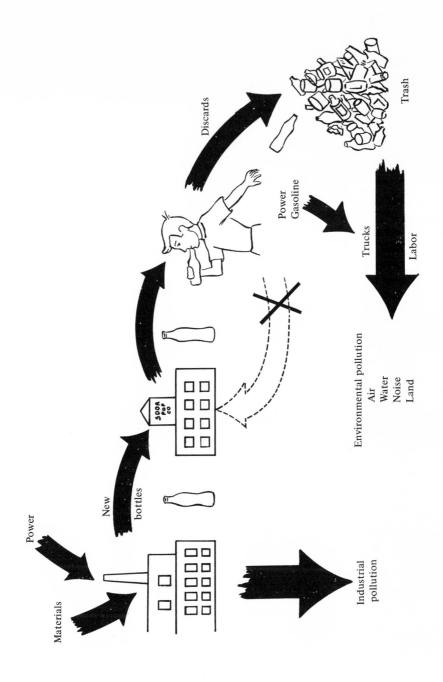

Materials

Power

New bottles

SODA POP CO

Discards

Trash

Power Gasoline

Trucks

Labor

Environmental pollution
Air
Water
Noise
Land

Industrial pollution

Contrasting Efficiencies
in Cyclic Processes

The two diagrams on page 173 contrast the efficiency with which calcium is recycled within a New England watershed and that with which iron and steel is recycled in U.S. industry. In the case of the watershed, 700 pounds of calcium are kept in continuous circulation through incorporation into living plants, return to soil by the decomposition of fallen leaves and rotting logs, and the decomposition of dead animals for each acre of forest. This circulation is maintained by an annual input of 12 pounds of calcium through the weathering of rocks and a correspondingly small loss in runoff of dissolved calcium salts. The iron and steel industry manages to recycle 50 million tons each year while adding a fresh input of 70 million tons. The input is balanced by an irrecoverable loss of 30 million tons of iron and steel and the "storage" of 40 million tons in steel bridges, buildings, ships, and automobiles.

Calcium cycle in watershed

Iron and steel "cycle" in U.S. industry

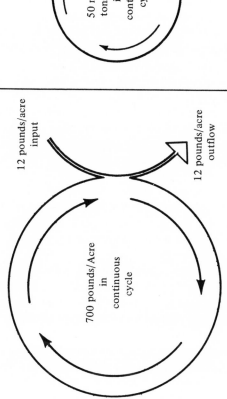

70 millions tons/year
fresh input

50 million
tons/year
in
continuous
cycle

70 million tons/year
outflow

= 30 million tons
lost
+
40 million tons
to "reservoir"

12 pounds/acre
input

12 pounds/acre
outflow

700 pounds/Acre
in
continuous
cycle

173

How to Think
About the Environment*

Max Ways

People who center their anxiety on "the population explosion" see the problem much too narrowly. In the U.S. and other advanced countries, population has been increasing less rapidly by far than the explosive acceleration of the total energy and total mass deployed. If the population declined and technology continued to breed, without any improvement in the arrangements for its prudent use, a small fraction of the present U.S. population could complete the destruction of the physical environment while jostling one another for room.

The casualties of a withdrawal from technology would be heavier than many suppose. Everybody, of course, has his own examples of unnecessary technologies, unnecessary products, unnecessary activities. But because we are, thank God, diverse in our wants, the lists do not agree. . . .We will not improve our environmental situation by recommending a technological retreat on the basis of what each of us considers the superfluous items in the households of his neighbors. . . .

Since we are not going to choose . . . a retreat from technology as a deliberate social policy, sheer practicality forces us to seek another way out. In that quest we have to ask seriously why the U.S. and all the other advanced countries have failed so dismally in handling the unwanted effects of technology. . . .

The Western tendency to objectify nature — to see it "from outside" — is undoubtedly responsible for much arrogant and insensitive handling of the material world. But it ought not be forgotten that this same attitudinal "separation" of man from nature forms the basis of man's ever increasing knowledge of nature. . . .

The scientific method began to separate one aspect of nature from another for purposes of study. This superlatively effective way of discovering solidly

*Mr. Ways is a member of Fortune's Board of Editors. This material originally appeared in the February 1970 issue of Fortune which has since been republished by Harper & Row as a book titled <u>The Environment: A National Mission for the Seventies.</u> © 1970 Time Inc.

verifiable truths tends, precisely because it is sharply focused, to ignore whatever lies outside its periphery of attention.

Science, seeking only to know, is guiltless of direct aggression against the environment. But technology, devoted to action, feeds ravenously upon the discoveries of science. Although its categories are not the same as those of science, technology in its own way is also highly specialized, directed toward narrowly defined aims. As its power rises, technology's "side effects," the consequences lying outside its tunneled field of purpose, proliferate with disasterous consequences to the environment — among other unintended victims. . . .

We have permitted the free combination of individuals on the basis of shared specific aims. By means of such groups, mainly corporations, we have organized and stimulated technological advance, matching techniques to particular group aims. Though this pattern all too often ignores the undesirable side effects of its singleminded thrusts, it fits . . . closely with the evolution toward human diversity and freedom . . . .

Though the principle of segregated attention proves gloriously successful— in research, in work, and in government — it can collide disastrously with the principle of unity. For each man is a unit though his skills and wants may be various. A society is a unit as well as a multitude. Nature, most marvelously connected throughout all its diversities, is a unit. Violation of these unities invites penalties and poses formidable tasks of re-integration.

Here we come to the root cause of our abuse of the environment: in modern society the principle of fragmentation, outrunning the principle of unity, is producing a higher and higher degree of disorder and disutility . . . .

The chief product of the future society is destined to be not food, not things, but the quality of the society itself. High on that list of what we mean by quality stands the question of how we deal with the material world, related as it is to how we deal with one another. That we have the wealth and the power to achieve a better environment is sure. That we will have the wisdom and charity to do so remains — and must always remain — uncertain.

Personal Commentary

Mr. Ways makes a number of excellent points in his article, some of which could not be reproduced here. In his effort, however, to show that the growth of technology, more than that of population, is responsible for the destruction of the environment, he may possibly have confused the issue by his use of conditional clauses. The population of the United States and of the world *is* growing; technology within the United States and other Western nations *is* expanding; technology *is* being exported to underdeveloped nations, and so the overall worldwide use of technology *is* expanding at a tremendous rate. The destruction of the environment is not a local problem; therefore, I contend that the numbers of persons *now living* multiplied by the amount of technology *now being made available to each* exceeds the carrying capacity of the earth's environment. "Ifs" and "coulds" tend to conceal this pressing issue.

No one has seriously suggested abandoning all of modern technology; the serious suggestions have advocated *control* – especially control of wasteful practices. Effective control, however, will necessitate lists of expendable items! If there are already too many persons for too much technology and if all persons are entitled to a fair share of the earth's resources, someone must sacrifice. The lists of items to be discarded may not agree (presumably they do not agree during wartime rationing either) and many persons may indeed vote to throw out their neighbors' goodies, but vote and throw out we must. Madison Avenue techniques and Madison Avenue cynicism must not be allowed to control this vote because, if the environmental crisis is to be met successfully, man must utilize all his rational thought.

Many of the points made by Mr. Ways parallel those made by Mr. McDermott.* For example, science ignores too often what is outside its periphery of attention; technology is too frequently directed toward narrowly defined aims and possesses a "tunneled field of purpose." McDermott also emphasized that certain acts seem rational from the technological point of view but appear insane to outsiders. I, in turn, have written of the need to enlarge the scope of various problems by pushing back their enclosing boundaries; only in this way can "externalities" be brought inside a problem where they belong.

In paragraphs not reproduced here, Mr. Ways speaks of the unique role of the university in meeting the current environmental crisis. This is one reason, of course, why I have written these essays. We can hope, for example, that the current crop of seniors leaving their colleges and universities to join "the system" will refuse to endorse or promote schemes that bring immediate profit but at tremendous future costs to society. One isolated rebel might be summarily dismissed by his employer as a misfit but an entire generation of young executives and engineers cannot be fired so easily. We can also imagine the notion of sabbatical leave spreading from academia to the executive arm of business and industry – periodic annual leaves (with pay) during which those who make decisions in industry can look around, talk to other types of persons, think, and ask themselves, "What is it that I am trying to make of this world?" At times my faith is badly shaken by the crude editorials I read in trade journals but I believe that fundamentally most persons want to preserve the world as a place fit for human habitation. And so when the chips are down and the lists are gathered up, I believe that most persons will agree on the sorts of sacrifices that must be made.

*See the essay entitled "Dilemmas Without Technical Solutions" in this volume.

Urban Sprawl:
A Splattering of People

A witness before a congressional committee described the growth of suburbs in these words: *

> A farm is sold and begins raising houses instead of potatoes – then another farm.
> Forests are cut, valleys are filled, streams are buried in storm sewers.
>   Traffic grows, roads are widened. Service stations . . . hamburger stands pockmark the highways. Traffic strangles. An expressway is cut through and brings cloverleafs which bring shopping centers. Relentlessly, the bits and pieces of a city are splattered across the landscape.
>   By this irrational process, non-communities are born – formless, without order, beauty or reason – with no visible respect for people or the land. Thousands of small, separate decisions – made with little or no relationship to one another nor to their composite impact – produce a major decision about the future of our cities and our civilization, a decision we have come to label "sprawl."

"Thousands of small, separate decisions – made with little relationship to one another. . . ." Again we encounter the "tragedy of the commons." Here is what has been called the "tyranny of small decisions." Each small decision is made on the basis of individual gain. Each decision is seemingly profitable when seen close up through the eyes of one or very few persons, but each contributes to the general decay of society when seen from afar. The flow of persons from the cities into the countryside, like the flow of a viscous fluid, has followed the path of least resistance, a path that has almost invariably been that which was most profitable to the real-estate developer.

Farmland is tillable land; tillable land requires little additional preparation when purchased for a suburban housing development. There are few trees; the rocks have long been removed; the land is reasonably level. Thus, the rural community learns that the old Bailey farm has been sold and that houses are going up. The full impact strikes home when school begins. Either the once-adequate school is overwhelmed the moment its doors open or (as was the case in Woodbury, Long Island, a number of years ago) a new family of children appears at the schoolhouse door every day of the school year as development houses are finished one by one and sold.

*The President's Council on Recreation and Natural Beauty, *From Sea to Shining Sea*, A Report on the American Environment – Our Natural Heritage (Washington, D.C.: Government Printing Office, 1968), p. 104.

Housing developments spread through the countryside, school district by school district, much as viruses infect a body, cell by cell. The first farm to go within a school district virtually dooms the remaining ones. The first housing development overcrowds the existing school facilities and calls for new construction and new staff. The increased taxes cause more farmers to sell to developers. Still more classroom space and teachers are needed. The additional taxes lead to the sale of more farmland until the school district has been "occupied." Farms have gone; Sylvan Hills, Belaire Homes, and Manor Estates have arrived.

What are some of the consequences of unplanned suburban development? Is urban sprawl something that can be tolerated for the moment in the hope that more rational solutions to population growth will be found in the future? These are difficult questions. In one sense, a plea for "taking it easy" can be interpreted (correctly in many instances) as an attitude equivalent to "I have mine, you wait for yours." Here again is Dr. Hardin's tragedy of the commons. On the other hand, the damage done by urban sprawl is in large measure irreversible; arable land once under a parking lot or any expressway is lost because the topsoil has been destroyed. This essay examines some long-lasting effects of the concerted move to suburbia. These effects are serious enough, in my opinion, to emphasize once more the need for eventual population control, the need to recognize that a land can hold only so many persons at a given level of comfort. Those of us who exist must be cared for in the best possible manner; nonexistent persons, the unborn, need not be invited wholesale to a largely imaginary "banquet of life."

Irresponsible behavior on the part of young blades was once described as "burning the candle at both ends." The acts of many societies could be described in identical terms because, in essence, the expression means thriving at the moment on reserves that will be needed in the future. The population of Ireland immediately prior to the potato famine expanded tremendously by reducing landholdings to a size that was inadequate for providing food in case the potato crop failed. The Irish expanded in number until they completely absorbed each year's crop of potatoes; when the crop did fail, starvation was the fate of many who could not flee abroad. The absorption of population growth by the preferential conversion of farmland into housing developments promises to repeat the mistakes made during the last century by the Irish. Less and less tillable land serves as the source of food for more and more persons. Heroic measures in the form of prodigious use of synthetic fertilizers, of an ever-increasing application of insecticides and herbicides, and of regional agricultural specialization supplemented by rural mechanization and a complex distribution network for farm produce manage to keep the growing population well-fed — for the moment. Let houses continue to be erected preferentially on farmland, however, so that dwindling numbers of acres support an exploding population, or let a calamity strike the food producers of the nation and we could easily find that we have sacrificed too much agricultural land too soon. Indeed, a nation that has no policy for population regulation would be well advised to treat agricultural land as an invaluable natural resource; to do otherwise is to burn both ends of a colossal candle.

The spread of houses into rural suburban areas does much more than turn prime farmland into basements and driveways. There are a host of enterprises that follow suburban development. First among these is the erection of one-story, multi-acred shopping plazas. Second is the hasty construction of highways that serve to link a motley assortment of residential areas. The land is literally covered by asphalt and cement. Because of the asphalt driveways, the new streets, and the enormous parking lots at shopping centers, at rennovated commuter stations, and at drive-in theaters, very little rain is permitted to enter the soil. From personal experience I know that the growth of one Long Island Rail Road parking lot from an informal arrangement accommodating a half-dozen automobiles to a (still inadequate) thousand-car parking lot caused enormous local floods after each rainstorm. Eventually the runoff caused the destruction of a nearby artificial lake; it exceeded the lake's capacity so that the lake spilled over and destroyed its earthen dam. This destruction occurred in an area where regulations concerning the construction of sumps for the preservation of Long Island's water table are unusually severe; nevertheless, large quantities of fresh water desperately needed by Long Island residents are lost by direct runoff into the sea.

The interplay of the inner city and its surrounding suburbs is exceedingly complex, but its overall direction is clear despite a multitude of kaleidoscopic minutiae. At any moment the opportunity to leave the city in order to live in the country is an attractive one. It is an attractive opportunity for the first persons to avail themselves of it. Because the departure of the early escapees lowers the quality of urban life, the opportunity to leave the city remains attractive for the second, third, and later contingents of escapees despite the lowered attractiveness of the suburbs themselves. The decline in quality of urban life caused by the departure of middle-class wage earners and their families, by the daily return of the commuter to the city by automobile, and the spread of small industries within the city make the prospect of escape to the suburbs a perennially appealing one for the remaining urban dwellers.

The distribution of the population throughout the countryside in single-family dwellings (a distribution that is seemingly encouraged by the Secretary of Housing and Urban Development) is a luxury the nation can ill afford. In the first place, as we have already seen, these dwellings tend to be erected on prime agricultural land because, for the moment, such land is cheap — cheap, that is, when one considers that its further development requires a relatively small investment. Thus, precious farmland is irretrievably lost. In the second place, the thin film of houses encourages a liberal sprinkling of neighborhood shopping plazas and other commercial areas, all of which, for the convenience of both the customers and suppliers, must be interconnected by means of arterial highways and heavily traveled avenues. These appurtenances require more prime land, which, once covered, is lost forever. In the third place, the need to move about — for the head of the family to drive to the city, for the mother to chauffeur young children, and, perhaps, for grown children to gain partial independence — requires an inordinate number of automobiles in suburbia. Here is the source of a great deal of the automobile exhaust pollution that engulfs urban areas — commuters, shoppers, and local deliveries. In the fourth place, the uncontrolled growth of suburban areas tends to destroy those very features that were initially attractive. Lakes and ponds become polluted by phosphate-containing detergents and sewage. Scenic hills are eventually bulldozed so that houses can be thrown up where they have no right to exist and, because the hills have been bulldozed, runoff from the hills floods older, nearby homes. Finally, the ill-advised exodus of families from the cities leaves large ·segments of once-residential urban areas at the mercy of manipulators, small-scale business, and shops specializing in peripheral merchandise such as cheap trinkets and auto parts. The city street as a center for an animated family and community life disappears; the result is the all too familiar inner-city decay.

To preserve our surrounding environment despite growing numbers of persons, to preserve our farmlands that will in time be needed to feed these persons, to reduce the need for private automobiles to a minimum, and, conversely, to encourage public mass transportation to the utmost, urban planners must turn to well-designed high-rise apartments or high-density urban centers rather than to a monomolecular film of single-family homes. Central Park in New York City contains 840 acres. Under rather crowded conditions some 2,000 families might be housed in single-family dwellings on this much land. Ten or 15 well-designed apartments could also accommodate 2,000 families on the same acreage and by careful design be spaced so that in addition the tenants would have luxuriously spacious shopping and recreational areas at their very doorsteps. The land, like the surface of a beautiful lake, can be destroyed by a film of pollutants; the same pollutants restrained as widely spaced globs are tolerable.

The Need to Solve
the Right Problem

In a number of essays I criticize shortsighted solutions to problems of pollution and of conserving natural resources. These solutions are posed by persons who fail to see (for some, the blindness is self-imposed) more than a part, rather than the entire, problem. These persons can be optimistic in the face of disaster; they can be expected to complain as matters begin to improve.

The following diagrams illustrate small and large views of the problem of the pollution of our nation's waterways by nitrogen and phosphorus. The first is the view expounded by an agronomist in attempting to absolve *the farmer* from blame. Runoff from fertilized fields might add nitrogen and phosphorus to nearby streams, thus causing their eutrophication and that of the lakes and rivers into which they feed. The agronomist, in his defense of farmers, points out that as much nitrogen and phosphorus are removed from the soil by the harvest as are added as fertilizer; consequently, nothing remains to run off. The farmer, according to this account, is not a polluter.

The second diagram includes more details than does the first, which is shown within the second by the dashed circle. It now appears that the balance between nitrogen and phosphorus added to the soil and that removed with the harvest has little or nothing to do with the larger problem of water pollution resulting from *modern agricultural practice*. The truth revealed by the larger diagram is that nitrogen and phosphorus are removed from air, natural gas, and phosphate ore only to pass down a one-way path into the streams, lakes, and rivers where they cause pollution. At no stage are these elements recycled because that would call for extensive use of manure and sewage as fertilizers; under our economic system, where environmental damage is not computed and natural resources are said to be infinite in quantity, recycling proves to be expensive.

The annual consumption of nitrogen and phosphorus in the nation's agriculture amounts to some 10 million tons of the former and 2 million tons of the latter. To obtain such large quantities of these two elements requires the equivalent of 400 billion kilowatt-hours of electrical energy, an amount equal to or slightly larger than that used by the nation's steel industry. This energy is used to run a gigantic wasteful pump, pumping nitrogen and phosphorus from the air and the land into the sea.

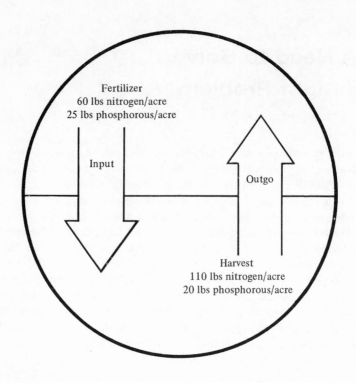

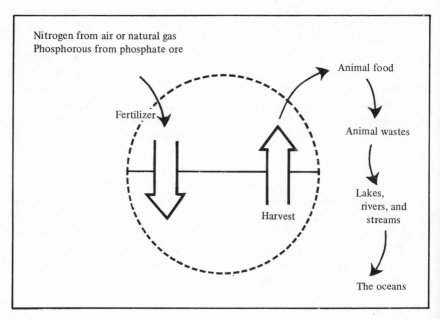

The account I have given here in connection with the two diagrams should be contrasted with a statement by Hans H. Landsberg:*

The small percentage of total energy that can be tagged as being consumed in farming is largely in the form of liquid fuels for driving trucks, tractors, and other farm machinery. . . . One can conclude that agriculture uses about 10 per cent of total consumption of liquid petroleum products. This, however, includes fuel consumed not only for productive purposes, but also for driving automobiles in the everyday pursuits of a normal household, for space heating, cooking, etc.

*Hans H. Landsberg, *Natural Resources for U.S. Growth* (Baltimore: Johns Hopkins Press, 1964), pp. 107-108.

Shadow and Substance

The theme of several essays in this collection has been that, because the earth is finite, man cannot for long continue to increase both his numbers and his demands on the environment. There is ample reason to believe even now (according to the accounts I have presented) that if Western-type industrial societies were to spread to all peoples, man's demands could not be met by our once-bountiful earth.

Not all persons agree with the pessimistic theme developed in these essays. I propose in the following pages to examine some of the reasons for these divergent views. Information that is especially valuable for identifying differences of opinion has been gleaned from two books prepared largely by and for economists.* Precisely because this material has been chosen in order to illustrate *differences* of opinion, the impression might arise that my views differ from those of economists on all matters. This impression would be false. Indeed, in an assessment of problems of water pollution and sewage disposal in suburbia, M. Mason Gaffney (B:95) writes that "on the sprawling urban fringe we are close enough to get in each other's way [pollution of water supply by septic tanks] and too far apart to do anything about it [excessive costs for suburban sewage systems]." Later, as if in response to a direct appeal from me, he adds, "We need a means for quick and complete transition of land from rural to urban density." Gaffney's opposition to the ever-spreading film of single family dwellings appears to be as intense as my own.**

Along with many fellow biologists, I believe that the environmental problem can be summarized as follows. Man is but one member of an interdependent community of life on earth. The earth — land, sea, and air — is finite; it can offer mankind only a limited supply of material resources and it can absorb only a limited amount of wastes. Man's goal in using resources is to utilize them in such a manner that they will last indefinitely — *in perpetuity.* Fifty years or 100 years are not "forever." Cyclic use of resources is the only pattern of use that avoids waste. Much of today's efficient technology is in truth based upon wastage of materials by their one-way flow through our economy: phosphate ore → phosphorus → detergent → sea; or, only slightly more devious, phosphate ore → phosphorus → fertilizer → plants → animals → organic wastes →

---

*(a) Hans H. Landsberg, *Natural Resources for U.S. Growth: A Look Ahead to the Year 2000* (Baltimore: Johns Hopkins Press, 1964), and (b) Henry Jarrett, ed., *Environmental Quality in a Growing Economy* (Baltimore: Johns Hopkins Press, 1966). [Page references to these books appear in the following format: (B:12).]

**See the essay entitled "Urban Sprawl: a Splattering of People" in this volume.

sea. Finally, the environmental problem is a tremendously serious one that will be solved neither by ad hoc measures nor by shortsighted calculations.

Before seriously examining the views expressed by economists, I must destroy two myths. There has been no serious recommendation by biologists that society revert to the horse-and-buggy days of a bygone era; biologists merely point out that the earth can yield only so much on a sustained basis and that mankind is now demanding that much or more. To an economist, a constant Gross National Product (GNP) forbodes cultural retrogression; the biologist has no comparable fear. Furthermore, no biologist to my knowledge makes "continued statements that water should be as clean as physically possible, that the air must never be used for waste disposal, that no bird or beast should ever be the victim of a pesticide. . . ." (B:71). I suspect that these alleged statements are the figments of someone's imagination; they may make a useful ploy to gain momentary advantage in a debate but they are not helpful in our search for lasting solutions to problems that *must* be solved.

In looking ahead to the year 2000, Hans H. Landsberg (A:239) acknowledges that he "shall not consider population among the major issues." Population growth is taken as "given" by the economist; he then tries to make the best of it. I do not take population growth as given because, in contrast to Landsberg, I am attempting to persuade persons to change their behavior; Landsberg tells them the likely consequences of their behavior as it exists.

The need to preserve our earth in perpetuity is based on a Golden-Rule-type of ethics: all persons are equal; my neighbor deserves the same breaks I demand for myself; our unborn grandchildren deserve the same as both my neighbor and I now claim for ourselves. Individuals living in the year 2000 deserve an earth as well-preserved as it is possible for us to keep it. This attitude is not universally accepted by economists nor, according to them, by persons generally. Everyone, they say, discounts the future – even their own. According to Kenneth E. Boulding (B:12) the future is discounted at a rate of 5 per cent to 10 per cent per annum. At 5 per cent, posterity's rates (in dollar values) halve every 14 years and each generation is "valued" at 25 per cent by the previous one; at a discount rate of 10 per cent, the value would be much less (4 per cent). Individuals of one generation will act to forestall a problem for the succeeding one only when it appears that the problem, if untended, would grow to be four or more times as large in the intervening years. This may have been a tolerable ethic in a frontier society but it is tolerable no longer. It is a fatal ethic in an age when the magnitude of problems doubles in size even as we attempt to understand them. Once more, my efforts are spent in attempting to change persons' attitudes. I cannot accept as "given" the selfish attitude: "What has posterity done for me?" "Forever" is a long time, much longer than the three or four decades economists are willing to stretch either their necks or their imaginations. But we *must* behave in such a manner that mankind and other life on earth will continue to exist forever. There is no acceptable rate at which the future may be discounted.

The depletion of resources and the gross contamination of the environment, in my opinion, are problems that will be solved only by recycling materials so that they are constantly in use, so that only the merest trickle "vanishes" into the waste heap. Recycling conserves materials and prevents waste. Not all economists have seen the problem as I have and, as a consequence, their proposed solutions appear to me to be false ones. One example (B:75) is the recommendation that one stream of several in an industrial area be reserved for waste disposal while the remainder are preserved for recreational use. This suggestion, unless it is coupled with other procedures such as local recycling, does not diminish the amount of either waste produced or resources consumed. Three hundred pounds of mercury would flow into Lake Erie each day whether it was discharged into Elk Creek, Mill Creek, and Conneaut Creek or into just one of these three.

A second, more blatant, example (B:96) is the suggestion that pollution can be solved by *importing* products whose manufacture or mining pollute the environment. This is a useless suggestion because we are discussing a worldwide problem; the contamination of the environment and the depletion of natural resources that concern us are not local problems. The suggestion is a cruel and insensitive one as well; the day when American dollars can be used to buy the health and lives of strangers is fast disappearing.

Environmental problems are too serious for offhand, optimistic suggestions; thoughtless persons too often assume that a *suggestion*, especially one in print, is an effective *solution*. Following the uproar over air pollution by automobile engines, one gasoline company announced in full-page advertisements that a secret ingredient was being added to its gasoline in order to prevent pollution. Economists also have their dreams; methane-producing algae is one of them (B:8). Surely to the sarcastic delight of McDermott,* the claim has been made (B:126) that the rising level of per capita production and the accelerating pace of technological change enlarge the conditions of choice in the direction of greater freedom. I see matters quite differently. In the past, technological advance coincided largely with wartime effort; the problems of war called for technological innovation. Today, technology itself gives rise to problems that call for additional innovation. We have entered a period of a self-sustained technological explosion that cannot be dampened by innovative procedures that raise problems more serious than those they solve.

Economists concern themselves with money rather than with material things, and for this reason much of what they say means something else than what a biologist would mean even if he were to use the same words. For example, an economist says (A:85) that an American may consume more paper in one trip to the supermarket than an Asian in several months. "Consume" means in this example that the dollar value of the paper vanishes during the

*See the essay entitled "Dilemmas Without Technical Solutions" in this volume.

transaction at the supermarket. In reality, the paper is not consumed; it goes into the trash barrel where, with luck, it will be carted to the city dump either to be burned or to be covered with a layer of dirt. The pollution of the environment is caused by physical substances, not by dollars, and physical substances merely pass through the hands of "consumers." Here is the basis for the constant harping on the recycling of materials; here, too, is my basis for contrasting "shadow" and "substance."

The Gross National Product furnishes the economist a measure of the nation's economic health. An exponentially expanding GNP is supposed to be good. It may expand because either productivity increases or dollar values change (inflation). With an increasing population, an increasing GNP is required merely to maintain a constant standard of living. A constant population would no longer require an increasing GNP in the sense, at least, of an increasing use of natural resources. Once more the economists differ in outlook from the biologists. Economists seem to regard an expanding GNP as intrinsically good; their attitude is by now instinctive. Landsberg, who concludes from his study that the natural resources of the United States will indeed be adequate for *our* needs in the year 2000, estimates that the GNP will double every 20 years for the rest of the century – *better* than we have done in recent years but *less well* than many would hope (A:19).

Much of the GNP consists of trinkets and services which, in my estimation, are foisted upon the public by industry (and their advertising agencies) rather than demanded by consumers, as economists would have us believe. Professor John K. Galbraith has argued that there is no such thing as "consumer demand," that, on the contrary, consumers buy what is made, advertised, and put before them. I also hold this view and, because I am disturbed by what appears to me as an unpardonable squandering of limited resources, I am appalled that "there are those who speak of 'annual-model' houses, like new-model automobiles" (A:45). I am also appalled at those who argue that success or failure under prevailing market conditions should determine the fate of any product or service, that railroads for example should disappear if they cannot meet today's demands for freight and passenger service. Such a shortsighted view virtually guarantees the success of those who squander natural resources most effectively and concern themselves the least with the discomfort and harm done by their noxious effluents.

In reading the books by Landsberg and Jarrett, I encountered many references to environmental quality. With the exception of Professor René Dubos, each author deals with the environment as something to be looked at or appreciated esthetically. Because they have not grasped the dangers arising from the destruction of the environment, the solutions these persons propose are cosmetic ones. Landsberg (A:248) says outright that the question of the natural environment is one of pleasant surroundings. Beauty, comfort, and health are some, but by no means all, of the environmental problem. The large aspect of the environment concerns the natural cycles that have been built into the system of life on earth over hundreds of millions of years and which have been virtually stable despite the evolutionary details that have occurred within the plant and animal kingdoms. The only massive exception that leaps to mind is the accumulation of atmospheric oxygen and of coal, oil, and natural gas as a result of the metabolic action of green plants. Man is dependent upon the continued functioning of these cycles. To destroy oxygen producers is to destroy the available supply of oxygen. To burn all carbon of organic origin is to remove all oxygen from the air. The destruction of denitrifying bacteria in an effort to enhance the role of liquid ammonia as a fertilizer would lead ultimately to a loss of nitrogen from the air. Small stockpiles of useful materials persist within a cyclic pathway only as long as the cycle operates; they arise in establishing equilibria between rates of input and rates of outgo. Break the cycle and the system stops with an enormous accumulation of the least useful substance.

Because of their overriding concern with the superficialities of environmental quality, economists often discuss at length problems that biologists might circumvent altogether. Dr. Ralph Turvey (B:49ff), for example, analyzes the conflict between farmers and the owners of brickworks in Britain. Fluorine that is emitted in the smoke from brickworks causes fluorosis in cattle. It causes the animals' teeth to mottle and wear faster than normal. Bones grow deformed and brittle. Milk production is reduced. The value of affected animals drops so that many must be slaughtered. The ensuing discussion touches on matters of prior occupation, on the relative merits of removing brickworks or farmers from a given area, and of coexistence under a system of economic compensation. Not touched upon in Dr. Turvey's discussion are questions that are worthy of a footnote even though their answers may already be known. What is the effect of fluorine emission on the farmers, their wives, and their children? What is the fate of fluorine inhaled and ingested by cattle? Does it pass into the milk? Into the meat? Does this contamination of dairy products, if it exists, represent a public-health hazard? In many respects, the purely economic aspects of the problem to which Dr. Turvey has addressed himself are trifling in comparison with those that bear on the health of the community.

Until now luck has been on the side of man; he has wrought relatively tremendous changes in his immediate surroundings but, except during the past two or three decades, these have been small disturbances when viewed on a global scale. Today matters have changed. Chemicals that are severely toxic in concentrations as low as 1 part in 10 or 100 millions are made in quantities as large as 100 million or 1 billion pounds *each year*; consequently, man has the capacity to affect in great but largely unknown ways all forms of life on earth. Mercuric compounds are among the most toxic known; the Food and Drug Administration considers 5 parts per 10 million in fish as dangerous contamination. Nevertheless, chemical firms have followed the economical practice of dumping hundreds of pounds of mercury daily into the nation's rivers and lakes; as a result, much more than the beauty of the environment is now threatened in many states. Such foolhardy "economics" endangers the continued existence of the network of life within which man exists *and only within which he will continue to exist.*

Economists are poor biologists (the reverse must be equally true). Quality of water seems to be equated solely with the amount of dissolved oxygen. This attitude leads in turn to the notion that a demand for high-quality streams can be met, if needed, by using aeration systems to add oxygen to the water. Only in conjunction with other suggestions does the futility of aeration become clear. In monetary terms, it is cheaper to squeeze clean drinking water from a highly polluted stream than it is to remove impurities at their source so that the stream itself is clean. Thus, "the need to prepare drinking water cannot justify particularly high standards of stream water quality" (B:74). Now, aeration of stream water in order to increase its oxygen content must be an acceptable practice only in the case of relatively clean water; in a rather thin soup of organic matter, aeration simply produces a rich bacterial culture that need not represent a substantial improvement in quality at all. In presenting this argument, I ignore lead, mercury, arsenic, phenols, and other toxic industrial and agricultural chemicals, which pose dangers that are unmitigated by oxygenation.

The main point of the preceding paragraphs has been that economists and their technological colleagues frequently propose solutions to environmental problems that upon close scrutiny prove not to be solutions at all. They deal with the shadow rather than the substance of the matter. These persons seem not to have grasped fully the extent of the problems before them. In paying lip service to the matter, they remind us that one way to get an idea is to get it all wrong. An economist, to be sure, must base his projections on the manner in which persons behave. There is no need, however, to assume that he can do no more and, consequently, either refrain from attempts to change people's ways or criticize such attempts when they are made by others.

People often act in illogical and irrational ways; educators try (or should try) to minimize the frequency of such irrational acts. Educated persons are, we hope, more likely than ignorant ones to make wise decisions. Gilbert F. White points out (B:108) that "early efforts to develop public water supplies in the United States encountered serious inertia because many people did not believe in the germ theory of disease." Education succeeded in this case; much of the teaching was done at the turn of the century in quaint, elementary-school hygiene courses.

If a general disbelief in the germ theory thwarted public water measures of a half-century or more ago, what sorts of disbeliefs will thwart efforts to slow down our modern technology? It is easy to point, as many sociologists and economists do, at the soft data upon which biologists base their predictions. I suspect, however, that there is a more profound disbelief. It rejects the idea that a world created especially for man can be destroyed by man's activities or even that man is part of nature. If this suspicion is correct, then the 100 years during which the word "evolution" could not be used in our public schools have set the stage for a monstrous catastrophe.

SECTION THREE

Ecology

Introduction

In recent years Ecology has gained a reputation as the subversive science.* At increasingly frequent intervals ecologists oppose actions that for a half-century or more have been looked upon as signs of progress. It has fallen to the ecologist to oppose the superhighway to the remote recreational area because in the past such highways unopposed have led to real-estate speculation, housing developments, shopping plazas, flimsy lunch counters, and the irretrievable loss of the original scenic attraction. It has fallen to the ecologist to fight the pollution of the environment because he more than others appreciates the delicate balances that keep the different species both in existence and in check; he alone appreciates the burden that certain inconspicuous species bear for man. It has fallen to the ecologist to fight industries intent upon exploiting the environment for the benefit of their stockholders and because, they say, what is good for industry must be good for the country. When the ecologist scores one of his rare victories, the industrialist joins the local politician in railing against these so-called experts and do-gooders.

We have encountered ecology before in discussing people and their needs — this was a discussion of "human ecology." We encountered it too in discussing the environment because ecology is literally the study of the "house" and the environment is man's house. Now we encounter it once more in an attempt to understand the forces that operate to maintain the diversity of life about us.

Three selections have been chosen for the vividness with which each of their authors has described the biological and physical aspects of a chosen environmental setting. The description of Las Marismas has been taken from *Iberia,* James A. Michener's massive love letter to Spain. Exceeding the thoroughness of many professional naturalists, Michener describes the panorama of life in Las Marismas through the four seasons of the year. *Os Sertoes* is the source of the description of the Caatingas, the arid backland of northeastern Brazil. Euclides da Cunha, the author of *Os Sertoes,* like Michener, takes us through the entire year, but in the Caatingas there are just two seasons — wet and dry — rather than four. The final selection, a description of a brackish lagoon, has been taken from *Venice* by James Morris, a historian who has been wholly captivated by the old city and its setting.

*This phrase has been borrowed from Paul Shepard and David McKinley, eds., *The Subversive Science: Essays Toward an Ecology of Man* (Boston: Houghton Mifflin Company, 1969).

In one passage Morris writes:

The lagoon can be an uncommonly lonely place. Its water, as the Venetians would say, is very wet. Its mud is horribly sticky. On many evenings, even in summer, a chill unfriendly wind blows up, making the water grey and choppy, and the horizons infinitely distant. Often, as you push your crippled boat laboriously across the flats, the mud gurgling around your legs, there is nothing to be seen but a solitary silent islet, a far-away rickety shack, the long dim line of the mainland, or the slow sails of the fishing-smacks which, loitering constantly in the water, give these waters some of their slow sad indolence. It is no use shouting. . . . The empty spaces are so wide, the high arch of the sky is so deadening, the wind so gusty, the tide so swift, that nowhere on earth could feel much lonelier, when you are stuck in the ooze of the Venetian lagoon.*

Unintentionally, Morris has described in this passage the birthplace of much of the harvest man reaps from the sea. Newly hatched fish need marsh grass in which to hide and water teeming with microscopic organisms on which to feed. Unfortunately, for too many persons Morris has also described here a convenient dumping ground for rubbish and other fill, a potential airport, or the site for future apartment houses and seashore restaurants.

*From *The World of Venice,* by James Morris. Copyright © 1960 by James Morris. Reprinted by permission of Pantheon Books, a division of Random House, Inc., and Faber and Faber, Ltd.

Las Marismas*

James A. Michener

One wintry day when storm clouds hung low over Sevilla, I set out on a journey to the south to complete a task I had set for myself some years earlier. I had been planning a novel with a Mexican setting, and because of the Spanish background of one of my characters, I required to know something of life on a Spanish ranch given over to the rearing of fighting bulls. I had selected as my prototype the historic and honored ranch of Concha y Sierra (Shell and Mountain Range) and a matador living in Sevilla had agreed to show it to me. He respected the Concha bulls because of their heroic performance during the past eighty years.

'The ranch lies in the swamps,' he warned me as we set forth, and this seemed an unlikely statement, since one visualizes a bull ranch as occupying hard, rough soil which strengthens the bull's legs. 'A common misunderstanding,' the matador assured me. 'It's the nature of the grass, the minerals in the water. . . something in the essence of the land and not its hardness. That's what makes a good bull. Anyway, the Concha y Sierras live in the swamps.'

We drove south to the end of the road. Parking our car in the rain, we set out on foot along a narrow earthen path that might have been passable in the dry season but which was now so muddy that going was difficult; if we stepped off the path on either side we were in swampland, not the green-covered stagnant swamp of fiction but an interminable area extending for miles in all directions, consisting of completely flat land covered with grass and two or three inches of water. 'These are the swamps of the Guadalquivir,' the matador said, and for the first time I looked out upon the infinite desolation which was to attract me so strongly and in so many ways.

The storm clouds swept rain squalls ahead of us which beat down upon the brackish waters to bounce back in tiny droplets, so that it seemed as if there were no horizon, as if sky and earth alike were made of mist and grayness. For two reasons that first day persists in memory. The immensity of the swamps astounded me; I had not realized that Spain included so large an area of primitive land, a retreat given over primarily to wildlife, where birds from all parts of Europe and Africa came in stupendous numbers to breed; this swamp, lying so close to Sevilla, was as wild as the seacoast of Iceland, as lonely as the steppes of Russia.

What I remember most vividly, however, is that on this introductory day, for reasons which no one has been able to explain, as the matador and I walked a flock of swallows stayed with us, perhaps a hundred in all, and when we took a step they swooped down to the tips of our shoes, then off into the sky, one after another, so that we moved in a kind of living mandorla such as encloses the saints in Italian religious painting. At times the swallows came within inches of our faces, swooping down with an exquisite grace past our fingertips and to our toes, flicking the swamp water with their wings. This continued for about half an hour, during which we seemed to be members of this agitated flock, participants in their spatial ballet, which moved with us wherever we went. It was one of the most charming experiences I have ever had in nature, comparable I suppose only to the day when I first skin-dived to the bottom of the coral beds off Hawaii; there was the same sense of kinesthetic beauty, of nature in motion, with me in the center and participating in the motion.

Why the swallows stayed with us for so long, I cannot guess; once during the walk I wondered if our feet, to which they seemed to be paying most attention, might be kicking up insects too small for us to see but inviting to the birds; but I had to dismiss this when I saw no evidence that the swallows were catching anything. There was also the possibility that our steps were throwing up droplets of water which the birds were taking in the air, but again there was no evidence that we were doing so, and I concluded that they were simply playing a game. This was not unreasonable, for obviously they were enjoying our walk as much as we were. At any rate, they served to remind me that these swamps existed as a realm for birds and that in entering it I was trespassing on their terrain.

The area is called Las Marismas (The Tidelands, in this instance a twofold tide, one coming in from the Atlantic Ocean direct and a more important one creeping up the Rio Guadalquivir to spread out over an immense area). Las Marismas is roughly forty miles from north to south, thirty-five miles from east to west, but it is not square and has an area of less than a thousand square miles, or about six hundred thousand acres, of which only about three hundred thousand could be called swampland proper, the rest being equally flat and bleak but free of water most of the years.

I was fortunate in visiting Las Marismas for the first time in winter, for this was the rainy season and I was thus able to see the bull ranch in maximum swamp condition; it seemed to me that about seventy percent of its land was either under water or was so water-logged that if I stepped on what appeared to be a solid tussock, it collapsed beneath me with a soft squish, so that my feet were again in water. It was on such land that the Concha y Sierra bulls flourished, but it was not until the matador led me to the dry area on which the ranch buildings stood, and I saw the famous brand of an S inside a C scrawled on the side of a corral, that I was ready to believe that this was the territory of the bulls about which I had read so much.

The Concha y Sierra bulls had a brave history, and many a noble head had gone from the bullring to the taxidermist's and from there to the wall of some museum, with a plaque beneath to inform the visitor as to what this bull had accomplished before he died.

On June 1, 1857, the Concha bull Barrabás participated in what the books describe as 'one of the most famous accidents in the history of bullfighting' in that, with a deft horn, it caught the full matador Manuel Dominguez under the chin and then in the right eye, gouging it out. It was assumed that Dominguez would die, for his face was laid open, but with a valor that had characterized his performance in the ring he survived, and three months later was fighting again as Spain's only one-eyed matador, having stipulated that for his return the bulls must again be from Concha y Sierra. For another seventeen years he fought with only one eye and enjoyed some of his best afternoons with Concha bulls. He is known in taurine history as Desperdicios (Cast-off Scraps, from the contemptuous manner in which he tossed aside his gouged-out eyeball).

On August 3, 1934, the Concha bull Hormigón (Concrete) verified his name by killing the beginning bullfighter Juan Jiménez in Valencia, and on May 18, 1941, in Madrid the gray Farolero (Lamplighter) killed the full matador Pascual Márquez, thus ending the career of a young fellow of great promise. On August 18, 1946, the Concha bull Jaranero (Carouser) killed young Eduardo Liceaga, brother of one of Mexico's best matadors. And so the story goes, with the great gray bulls of Concha y Sierra defending themselves valiantly in all plazas.

The matador and I left the ranch buildings and on horseback set off across the marshes to see if we could find any of the bulls in pasture, and after we had ridden for some time in the direction of the Guadalquivir, that meandering, desultory river of such force and quietness, the matador suddenly cried, 'Look!' and off to our left, rising from the reeds and thistles like an apparition, loomed a gray bull, his horns uptilted, his ears alert. We stopped the horses. He stopped. We stared at each other for several minutes, and then we saw, gradually appearing from the mists behind him, the shapes of fifteen or twenty other bulls, and slowly they moved toward us, not in anger but rather to see what we were doing. They came fairly close, much too close for me, but the matador said, 'They won't charge as long as they're together and we stay on the horses,' and so we stayed, among the bulls who had materialized from the swamps, and after a while they gradually drifted away and the mists enveloped them.

It was while wandering in this fashion in Las Marismas that I became aware of the Spanish seasons — the rain and the drought, the cold and the heat, the flowering and the harvest — and I decided that if I ever wrote about Spain, I would endeavor to cover each of the seasons. I have never spent all of any one calendar year in Spain, but I have visited each major area except Barcelona during at least two different seasons so as to see the effect of the passing year upon it; so far as I can remember, the only month I have missed is February, and it is possible that one of my Easter visits started in late February, but if so, it could have involved only a few days, and I do not remember it. I have seen Las Marismas in all seasons, never as much as I would have liked but enough to teach me a few facts about the land of Spain.

Spain! It hangs like a drying ox hide outside the southern door of Europe proper. Some have seen in its outlines the head of a knight encased in armor, the top of his casque in the Pyrenees, the tip of his chin at Portugal's Cabo de São Vicente, his nose at Lisboa, his iron-girt eyes looking westward across the Atlantic. I see Spain as a kaleidoscope of high, sun-baked plateaus, snow-crowned mountains and swamps of the Guadalquivir. No one of these images takes precedence over the other, for I have known fine days in each of these three contrasting terrains. That snow should be a permanent part of my image may surprise some, but Spain has very high mountains and even in the hottest part of July and August, when the plains literally crack open from the heat and when the blazing skies described in Spanish fiction hang everywhere, snow lies on the hills less than thirty miles north of Madrid. In the middle of August not long ago I drove across the mountains that separate the Bay of Biscay from the city of León and saw snow about me for mile after mile; at another time, when I had come to Spain in midsummer for my health, doctors in Philadelphia asked, 'Is it wise to visit Spain in July? Won't you suffocate?' But my plans took me to the high north, where during much of the summer I needed a topcoat at night.

But now we are speaking of Las Marismas, in the heart of the southland, and to appreciate it, and the land of Spain in general, we must watch it through one whole year, and if in doing so I seem to be speaking primarily of birds, that is appropriate, for here is one of the great bird sanctuaries of the world, as if nature, realizing how difficult it was going to be for birds to exist in a constantly encroaching world, had set aside this random swamp for their protection.

Winter

The Guadalquivir itself never freezes, of course, but occasionally areas of shallow water with low salt content will freeze to a thickness that would support a bird but not an animal. In winter the tides run full, and even if no additional rain fell they would be sufficient to fill the streams that crisscross the swamps; but rains do fall, most abundantly week after week, and the tides are reinforced, so that water stands on the land. The sky is mostly gray, but when storms have moved away, it becomes a royal blue set off by towering cumulus clouds moving in from the Atlantic; Las Marismas is a product of the ocean; it is not a somnolent offspring of the Mediterranean Sea.

Maximum flooding occurs in January, and then one can move by boat across large segments of the land, and in this boggy, completely flat wilderness life undergoes conspicuous modification. Animals like the deer and lynx have taken refuge by moving to preselected higher ground, and in their place come several million birds from northern Europe to prepare for breeding. Three hundred thousand game ducks have moved in to winter, ten thousand large geese. In January untold numbers of coots arrive to breed. Through a hundred centuries they have found in Las Marismas a plentiful food supply, for the marsh grasses provide seeds and roots and the shallow waters teem with swimming insects.

Toward the end of winter the birds begin to settle upon specific clumps of grass, testing them for strength and protection, and in March they begin weaving nests which will house them for the important work ahead in spring. Where do they get the material for the million or more nests they must build? From all imaginable sources but particularly from the weeds themselves. However, in the nests I have seen feathers, bits of mouse fur and even strands of hair from cattle.

In the winter men leave Las Marismas pretty much alone. Along its edges, of course, towns have grown up and in them live skilled hunters who know all the footpaths that cross the swamps. Here also are herdsmen who pasture their cattle on the grassy portions in the summer, and fishermen who work the Guadalquivir. And there are, I am glad to say, a handful of men who simply love the bleakness of the swamps and study it year after year, as they would a book, but in winter even they stay mostly at home,

Over the vast area one sees mostly the movement of birds, thousands upon thousands of them, birds that have known Siberia and the most remote fjords of Norway, that will spend their summers on the moors of Scotland and in the forests of Germany. They live in tremendous families, each associating with its own kind, but a single area of marshland may contain fifty different species, waiting through the long winter till their summer feeding places in the north have thawed.

On the thirteenth day of January each year, in obedience to one of those unfathomable rules that govern birds, storks fly north from Africa to their chimneys of Holland and Germany and continue to do so for some weeks, so that the Spanish have a saying which could be translated as:

At the day of St. Blas
The storks do pass.

Why they go north in midwinter has never been explained, but in midsummer they will go south, as if their calendar were askew.

But there are also, in these months, the birds that live permanently in the Spanish swamps, and in some ways these are the most interesting because they are the ones that we shall see in all seasons, like old friends. There is no more beautiful small bird in Europe than the goldfinch of Las Marismas, a tiny gem of color and design. I have watched a group of goldfinches for an entire morning and have never tired of their display, the flash of their color against the brown swamp, the chattering of their family life. Large numbers are trapped here and sold throughout Europe, for they make fine pets, and whenever I saw them caged in other parts of Spain, I thought of Las Marismas, for they seemed to take the swamps with them.

At the opposite end, so far as size is concerned, was quite another bird. I remember one day, when I was on the Atlantic Ocean edge of the swamps, seeing a huge creature fly into the crown of a tree. It was slightly smaller than a griffon vulture, which are common throughout Las Marismas, and of a different character. Since it remained motionless in the tree, I was able to study it at leisure, but it was not a bird with which I was familiar; later I learned that I had seen an imperial eagle, the noblest inhabitant of the swamps. There are partridge, too, and magpies, and crested coots, and purple gallinules, and a species of owl.

The resident bird which dominates the scene in winter is the cattle egret, a snowy-white bird with yellow legs, a long yellowish bill and a silhouette much like a heron's or a small stork's. They get their name from their habit of feeding not only with cattle but on them, so that if you are wandering through Las Marismas it is not unlikely that you will see a sleek and coiffured little egret riding like a debutante between the horns of some massive fighting bull as he grazes in the swampland, and I have often watched a herd of bulls and a flock of egrets as they blended together in such harmony that one would have thought they had been created as halves of a symbiosis. Certainly they form one of the most attractive features of the year. Regardless of where one sees them, the egrets are winsome birds, delicate in motion on the land and unforgettable in their broad-winged flight. They range far from the swamps and can often be seen in the fields near Sevilla, looking for insects, but no matter where they spend their days, at night they return to the swamps in flocks that number in the hundreds. They can be seen in all seasons but are most appreciated in winter, when they have least competition, as the total bird population is then at its smallest.

Primarily, winter in Las Marismas is a resting time, for the birds, for the animals, for the seed plants and for the men; but to see the swamps in this season is an intellectual challenge. Can you imagine what they will look like in summer? I failed the test, for I was unable to visualize this watery world, this endless waste of tussock and salt, becoming other than what it then was. I could not imagine the tranformation it was to undergo.

Spring

The rains cease. Evaporation begins, and with each inch that the water falls, grass springs up to take its place. What had seemed, only a few weeks ago, seventy-percent water, now seems ninety-percent grassy meadowland, but if one steps off established paths, he sinks in up to his knees, for the underlying water will remain until well into June. As the waters recede, the swamps cease being attractive to ducks and geese, who fly north in huge flocks to the thawing lakes of Russia; but as seed grasses appear, with their assurance of food, large numbers of terns and coots arrive to set up housekeeping, and the men of Sanlúcar de Barrameda, the town at the mouth of the Guadalquivir, prepare their boats for one of the strangest harvests in Europe.

It begins when grass has fairly well covered the swamps, so that horses and cows can be let in to forage. They eat such grass as shows above the water, in which they stand up to their knees, but as they feed they leave behind shreds of grass that float on the surface and these the terns and coots collect for their nests, which they build on circular flat constructions that float on water and stretch in all directions for miles.

Now comes the harvest. In these semi-floating nests the birds lay large quantities of eggs, and for as long as men have lived along Las Marismas they have poled their flat-bottomed boats into the marsh in spring, collecting these birds' eggs. They work in teams of six or seven men to a large boat, from which small skiffs, each bearing a single man, set out to explore tiny rivulets, gathering eggs which will later be sold for food. Year after year they rob the nests of hundreds of thousands of eggs, but the bird population seems not to suffer, for enough eggs are overlooked to ensure the perpetuation of the species.

Only once did I see a team of egg collectors in action. They came, as usual, from Sanlúcar, one of my favorite towns in Spain, a sun-baked, miserable dump of a place that looks much as it did in the days of Columbus and Magellan, who knew it well, a most authentic remnant of old Spain. Five men entered the area in a large boat painted blue and boasting an outboard motor. In the boat they carried a small skiff which they launched in the marshes. The man who poled it through the shallow waters was indefatigable, for he moved swiftly from one floating nest to the next, scooping up enormous numbers of eggs. How many did he gather in the short time I saw him? Probably five or six hundred, and he had touched only a small portion of those available to him. The men in the blue boat were drinking wine and encouraging him, and I never understood what the division of labor was supposed to be. Perhaps the man in the skiff was the only worker; the others may have come from Sanlúcar for the ride.

As the grasses grow and the land begins to solidify, small land birds begin to crowd Las Marismas in flocks of such magnitude that most Americans have no experience with which to compare them. Many arrive from Africa and the Holy

Land, and I shall never forget my astonishment, one spring day when I had arranged a picnic in Las Marismas for a group of friends, at seeing, near the clearing in which we ate, two of my favorite birds from Israel, the long-billed, inquisitive hoopoe and the brightly colored bee-eater. 'Are they native here?' I asked an expert who was sharing our picnic. 'No, they migrate from Africa but they arrive so regularly each year that we think of them as native.'

Even in spring, when the swamps have begun to look like land, it is the water birds that one remembers best, for now the avocets arrive, those delicate, long-legged birds with the upturned bills; I had not known the avocet until I spent some time in Colorado, where they were common, but the Spanish ones seem larger and more colorful. The stilts come now, too, and the slender-billed terns, so that what lakes remain are crowded with fascinating life, even though the spectacular ducks and geese have gone.

Summer

Summer is something to see in Las Marismas! Even though storm clouds occasionally hang over the Atlantic, the sky over the land becomes an incandescent arc producing temperatures that go well above a hundred degrees in the shade, if any can be found. Day after day the sky hangs there, motionless, relentless, drying up the waters and bringing the grasses to seed. What few streams remain are covered with golden pollen, and even their banks are barren for yards on each side. Young birds are everywhere, feeding on fallen seeds and slapping their awkward feet on the baked earth as they look for water. Jack rabbits appear in large numbers; they attract fox and lynx, who hunt them constantly, but it is from an unexpected source that the food supply becomes abundant.

As the accidental streams that crisscross Las Marismas dry up, multitudes of fat carp search frantically for the permanent rivers which will sustain them through the summer, and in great numbers they move in obedience to faulty instinct from one evaporating fragment of water to the next, until at last they perish in vast numbers on dry earth. At times their glistening bodies completely cover what had lately been a lake, but before they have a chance to rot and thus contaminate Las Marismas, flocks of kites and vultures, sensing the impending tragedy while flying over North Africa, swoop in and help the local birds clean up the carcasses.

The bird that seems to represent summer at its best is the heron. The large white ones appear in flocks of up to six thousand at one time, the smaller in flocks of twenty thousand, ranging over the entire area in white dignity. How can so many birds find food? They eat fish when they find them, and frogs, lizards and the larger insects. They scour the dried earth for remnants of the carp and uncover so much food that they prosper where other birds would fail.

A bizarre tragedy now occurs and one that I would have thought improbable had I not seen it. Among the hordes of aquatic birds that resided here in the spring, and I am speaking not of hundreds but of hundreds of thousands, most have left, but there are some who nested here, and they seem unable to believe that these watery lands are going to dry up, so in spite of mounting evidence in late June and early July, they linger on. Now the remorseless drought of late summer catches up with them and for some weeks the three-month-old ducklings search frantically for ponds which they knew existed in a given spot only a month before, but they find only sun-baked earth. Sometimes they march on webbed feet, three or four thousand in a small area, searching vainly for water, and one by one they perish.

Now the raptores move in on silent wings to kill off the survivors. The sharp-eared lynx darts out from his hiding place to catch his supper, while the fox and the rat keep watch. The mournful pilgrimage continues for the better part of a week, this noisy march of hopeless ducks trying to find water, and then Las Marismas is silent once more.

The extraordinary thing about this season is that in drying, the once-muddy areas of land become a perfect highway for automobiles — flat, even, undisturbed and so hard that cars throw no dust. I have several times driven far out into the summer swamps at thirty miles an hour, and when something interesting loomed ahead, at forty or even forty-five, and in this way have covered twenty or thirty miles with no inconvenience but with a sense of flying low in an airplane over a placid bay. Of course, the driver must have some general knowledge of where the permanent waterways are, for even if the water has evaporated, as is sometimes the case, the vanished rivulet leaves such a depression that the car could not cross it. Except for this limitation one can ride for hours across Las Marismas and see the skeletons of carp.

If one were to see Las Marismas for the first time in midsummer he would find it difficult to believe that the place should be called a swampland, for there is certainly no evidence to justify such a name. Perhaps marshland would be a better translation of the Spanish, or even the Scottish moorland because when dry, Las Marismas has many characteristics of the latter; but considering the area as it exists throughout four seasons, swampland is not an inappropriate description.

In summer many men come into Las Marismas, some to tend cattle, others to hunt and still others to wander through the wilderness as their ancestors have done for generations. The immense expanse of sky and the weirdness of the absolutely flat landscape exert a powerful appeal to these men, and one of their delights is to shoot a rabbit, skin it and then spread-eagle it on a structure made of three sticks tied together in the form of a Cross of Lorraine. The upright member of the cross is left long, so that it can be used a a handle for holding the rabbit over a fire of hot coals until the meat is hard and crisp. Salt is rubbed on the finished meat, which is cut into thin strips and mixed with raw tomatoes, peppers, much onion, garlic, olive oil and vinegar. 'Maybe the best salad a man can eat,' those who live along Las Marismas claim. For as long as men can remember, huntsmen who prowl the swamps have been entitled to shoot all the rabbits they need for food; the most recent estimate is that about eight thousand are taken each year.

But as the knowing men cross the hard-baked swamp they are careful to watch out for a menace which through the years has taken the lives of many animals and occasionally even of men. This is the ever-present ojo (eye), which stands invitingly here and there in attractive spots, a kind of minute oasis with a central eye of water, perhaps a swampy spring or well, and surrounding green grass and shrubs and sometimes even small trees. On the great arid swamp these ojos are most tempting, for they promise both water and shade, but they are treacherous because they also contain quicksand of a most virulent sort, and once it grabs hold of a leg it rarely lets go. Domestic animals wander into the ojo alone, get stuck and never break loose. If they die, they do so beside the carcass of some boar or deer that got stuck in exactly the same way a week before. Within a few hours the bones are white; vultures keep watch on the ojos.

No matter how well one knows Las Marismas he occasionally meets with surprises. One day as I was riding past a section of the swamp I saw long rows of what looked to be human beings, each bent forward from the waist as if gleaning a field for some lost object. I stopped and crossed the intervening land to see what they were doing; much of the land was under water but ridges had been left as footpaths, and after I had walked along these for a few hundred yards I saw that the bent-over people were women, with heavy nets over their heads and faces, and that they were engaged in transplanting rice, digging handfuls of young rice plants from the seed bed, where they grew in close profusion, and carrying them to the larger fields where they would be transplanted, one stalk at a time in the mud. The women were thus required to stand in water and bend over the soggy fields for eight and ten hours at a time, exactly as other women were doing in Asia.

The nets over the face served two purposes. If a woman bent close to the water on a sunny day for extended periods, the reflected rays of the sun would bounce up at her face and produce a sunburn that might in time cause cancer. More immediately, the nets kept away the hordes of mosquitoes that infested Las Marismas in summer, making it at times almost unbearable. 'If you're going into the swamps,' Spaniards told me repeatedly, 'take along some 612.' This was a potent insect-repellent that worked.

And occasionally as one penetrates the swamps he sees on the horizon a strange brown animal larger than a bull and thinks for the moment that his eyes are deceiving him. Then the animal moves, in an undulating manner, stops, twists his long neck and raises his long-nosed face. It's a camel. His ancestors were brought over from Africa in the latter part of the eighteenth century for use among the sand dunes around Sanlúcar; they adapted well to Spain, but peasants protested that they frightened them and that if God had wanted such ungainly beasts on Spanish soil He would have seen to the matter. Men interested in working the area tried to explain that the camels were harmless, but to no avail. In 1828 all those in the Sanlúcar area were rounded up, transported across the Guadalquivir and set free. There cannot be many left, and I suppose that within another decade they will have disappeared.

The kangaroos that were introduced somewhat later than the camels have already vanished, as have the monkeys which were brought to Sanlúcar from nearby Gibraltar. So far as climate and food are concerned, there is no reason why monkeys should not have prospered in Las Marismas, but once more the peasants of Sanlúcar, who must have been an unusually suspicious lot, protested that the almost-human faces of the monkeys scared them at night, and that if God had wanted such beasts. . . . The last monkeys were killed off about fifty years ago.

More rewarding than the camels that one occasionally sees are the melons, which are among Spain's best. They grow luxuriantly wherever sandy soil remains soft enough during the early summer to permit the vines to mature; most often the blazing sun absorbs all moisture and the plants wither, but if they survive, the fruit they yield is delicious. Apart from the rice, it is the only edible thing grown here commercially.

But whatever the season, Las Marismas is primarily the residence of birds, and what happens to men or camels or melons is a secondary concern, and so as summer ends one looks again to the sky and sees aloft the great bustards, accompanied by their cousins the little bustards, coming in to glean the hard ground for seeds and bees and insects. They fly in splendid circles and land in two or three hops. Their quick eyes scan an area in seconds to determine where the good feeding will be, and they pick the land clean, quarreling among themselves as to who saw which first. When they take their short hops and rise again into the air they see below them only a parched earth, blazing in heat as great as that of a desert, with the somnolent Guadalquivir wandering southward through the middle of Las Marismas, and at its mouth the sunburnt adobe of Sanlúcar de Barrameda, blistering in the sun as it did in the days when Columbus stopped there.

Autumn

The best time of year, for me, in Las Marismas is autumn, because three things happen to make it both bearable and exciting: the dreadful heat diminishes, so that temperatures become quite pleasing; the first rains come and with them a new color in all living things; and the movement of birds is captivating. As for the heat, Las Marismas, which is midsummer seems as hot as tropical Africa, stands at about the same latitude as San Francisco (Sevilla 37°, 27 N; San Francisco 37°, 40 ; Richmond 37°, 30 , Wichita 37°, 48). Therefore, when autumn comes and the baleful effect of winds blowing in from Africa and the Mediteranean has gone, the temperature is delightful; one wears loose clothing during the day, a substantial jacket of some kind after dark, and if one wishes to ride over Las Marismas at midnight to spot wildlife, a sturdy coat. I would suppose that anyone who loved the outdoors, and especially the tracking of birds in their larger migratory movements, would find Las Marismas in autumn almost irresistible, for then nature changes its aspect daily, and what has been a barren wasteland marked by whitened carcasses becomes a meadow which will sustain millions of birds in their migrations.

The change begins with the first rain. Year after year it arrives sometime between the twentieth and twenty-fifth of September, as if it had been waiting for summer officially to end. This is only a slight rain, not even enough to heal the cracks that mar the land, but it is followed in desultory fashion by one or two others. On Columbus Day, October 12, celebrated throughout the Spanish world as El dia de la raza (The Day of the Race), the people who live along the edges of Las Marismas enjoy their last guaranteed clear day and their picnics are apt to be gay, for with strange regularity, on October 13, comes the first drenching rain of sufficient duration to soak the ground, but even though enormous quantities of water fall in this and subsequent storms, there is still not enough for any to collect. No lakes re-form and the permanent rivers are no higher than they were before; this water seeps into the dried earth. In doing so it reactivates plants, and even before the swamps re-form they look as they did when water was plentiful.

Now a few courageous ducks and geese begin to arrive from Scandinavia, and they must be sorely frustrated by what they find, for there are no lakes and food is bitterly scarce, for seeds of autumn have not yet fallen and the water plants on whose roots the birds exist have not matured. There is even trouble in finding a lake on which to rest; most are dried basins, their cracks just beginning to heal. And even when some accidental lake is found, its water is extremely brackish and unable to provide the swimming life which ducks and geese use to supplant the seeds and roots. But these first arrivals struggle with their problems and no living thing in Las Marismas must welcome subsequent rains with more excitement. One day I watched as a group of land-bound geese wiggled and cried with delight as rain came down upon them; it was the promise of a fruitful autumn.

These newcomers face an additional problem, for when they are kept from their normal feeding and hiding grounds, they lay themselves open to attacks by the imperial eagles, who now move in for easy kills. Perhaps easy is the wrong word, because the eagles have to exercise real skill if they want to catch a graylag goose who has protected itself in the north for the last six months. No eagle flying alone has ever been seen to take a goose except by sheer accident, for although the eagle is stronger and has powerful talons, he cannot overtake a goose in full flight; pursuit is useless. Therefore, the eagle finds himself a partner and as a pair they become formidable. One flies rather high, in the fly-space of the goose, and somewhat awkwardly, so that the target gets the idea that he can outfly this enemy. The other eagle flies low and well behind the first, and as the awkward eagle maintains altitude on the goose and makes a series of futile passes at him, the big bird takes the easy way out and with adept spirals evades the eagle by dropping to a lower altitude, where the second eagle sweeps in with terrifying talons.

There is other death in Las Marismas now. The cattle who have been browsing all summer on the safe flat lands begin to withdraw to higher ground as the rains start to engulf them, and by following paths long established, they retreat, but often one stumbles into an ojo, now camouflaged by growing grass and shrubs, or he waits too long and is trapped on an islet, where he dies, or the long trek weakens him. In any case, he is carefully watched by the vultures who scout the vast expanses day after day. In Las Marismas nothing rots.

For human beings in the region the autumn is as exciting as it is for the birds and animals, because this is the season of the vendimia (vintage) when the first fruit of the vine is pressed to the accompaniment of week-long celebrations. If one wanted a single painting of Spain to remind him of the best of the country, he could do worse than choose Goya's exquisite painting of the vendimia, now hanging in the Prado in Madrid, in which idealized peasants bring in the grapes while a nobleman, his pretty wife and their little boy, dressed in green velvet and red sash, taste them. Spaniards love this unpretentious work, for it speaks to them of the land, the rich, hard land of southern Spain when the harvest is under way.

In Sanlúcar the vendimia is celebrated with the same rustic vigor that it has been for the last thousand years, but at nearby Jerez de la Frontera from which sherry takes its name, occurs the most renowned vendimia. Then the world-famous families who make and sell sherry — Domecq González, Byass, Osborne — set up kiosks where wine is served. Countrymen arrive to promenade in carriages drawn by six horses. There are bullfights and celebrations that last through the night. All around the rim of Las Marismas there is festivity in which Catholic Spain remembers pagan rituals and combines the old religion and the new in fascinating juxtapositions.

Now, too, is the time when huntsmen concentrate on the swamps, for the latter contain two enormous herds of deer, an indigenous red deer with pointed horns, which is held to be an honorable target for the huntsman and the grosser-formed fallow deer with palmated horns imported from Asia in the early 1900s, and not allowable as quarry for a gentleman. One autumn I was in a car filled with huntsmen speeding over the macadam-hard swampland, scouting for deer, and because I wear rather strong glasses I could see farther than my companions. 'Buck!' I shouted with some excitement as I spotted a handsome animal with large horns off in the distance. The car slowed down; the men looked; and there was silence as we drove on. I concluded that I had mistaken a doe for a buck, but shortly thereafter I spotted what could only be a buck. To my eyes he was majestic, with a spread of antler exceeding any I had seen before. 'Buck!' I cried, this time with firmness. The car stopped; the men looked; and in embarrassed silence drove on. On the third spotting, for I was still seeing animals before the others, I cried, 'Goddamn it, that's a buck.' This time the car did not even bother to slow down, but one of the Spanish gentlemen did whisper, 'Michener, look at the horns! No gentleman would shoot a beast like that.' I had been spotting fallow deer, and they didn't count. After a long silence I saw a herd of perhaps sixty deer, and they were different, red instead of spotted brown, pointed horns instead of palmate. 'Buck!' I shouted for the fourth time, and there stood a series of noble beasts with proper horns. My alarm caused some excitement, and the gentleman at my side whispered, 'Well done. Those are deer.'

I am not a huntsman, except with camera, so to me the deer-stalking of autumn was less exciting than the subtle transformation of the land. I have never seen Las Marismas in late autumn when the water system of winter is fairly well formed, but I have seen it twice in early autumn when the rains have begun to take effect, and to see lakes quietly come into being, to watch dead rivers creep back to life and above all to see the surface of the land begin to collect its water and soften itself from concrete into mud, with grasses and flowers gently appearing, is a profound experience. I wouldn't be able to say when the swamps had fully reestablished themselves, perhaps by the first weeks in December, but at any stage in the process they afford an insight into nature that one cannot obtain elsewhere. How beautiful this transformation is, how simple: the land was barren and a raceway for the wind; it is now a meadow and a home for birds.

As always, it is the birds that inspire. A cold wind comes down from Madrid and next day all the migrants from Africa have taken flight. Remember, it is hardly five hundred miles from Las Marismas to the first deserts of Africa, and beyond them it is a couple of days, flying time to the warmer regions in which the birds are accustomed to spend their winters. The bee-eaters, the hoopoe birds and some of the egrets depart, and in their place come the robin and woodcock and widgeon. For a short period the swampland seems relatively depopulated, for the birds that have fled were conspicuous in size and color whereas the newcomers are markedly less brilliant, but after a few weeks of emptiness the damage is repaired, for now the real flocks of ducks begin to appear, so that a lake that was empty one day may have a thousand birds the next, and as the waters replenish themselves, the birds do likewise, and the poetic year of Las Marismas draws to a close.

Rebellion in the Backlands

Euclides da Cunha

The Caatingas*

The traversing of the backland trails is then more exhausting than that of a barren steppe. In the latter case, the traveler at least has the relief of a broad horizon and free-sweeping plains.

The caatinga, on the other hand, stifles him; it cuts short his view, strikes him in the face, so to speak, and stuns him, enmeshes him in its spiny woof, and holds out no compensating attractions. It repulses him with its thorns and prickly leaves, its twigs sharp as lances; and it stretches out in front of him, for mile on mile, unchanging in its desolate aspect of leafless trees, of dried and twisted boughs, a turbulent maze of vegetation standing rigidly in space or spreading out sinuously along the ground, representing, as it would seem, the agonized struggles of a tortured, writhing flora.

This flora is one that does not show even the reduced number of species common to desert regions — stunted mimosas or rugged euphorbia over a carpeting of withered grasses. It is not replete with distinct vegetable varieties; its trees, viewed in conjunction, appear to be all of one family, with few genera, being confined almost to one unvarying species, barely differing in size and all having the same conformation, the same appearance of moribund vegetable growths, practically without trunks and with branches that start at the ground. Through an explicable effect of adaptation to the cramped conditions of an unfavorable environment, those very growths which in the forests are so diversified are here all fashioned in the same mold. They undergo a transformation and by a process of slow metamorphosis tend to an extremely limited number of types, characterized by those attributes which offer the greatest capacity for resistance.

This is a hard-and-fast rule. The struggle for existence which in the forests takes the form of an irrepressible tendency toward the light, with distended and elastic creepers that flee the stifling shade and, climbing aloft, cling rather to the sun's rays than to the trunks of the centuries-old trees — all this is here reversed; all here is more mysterious, more original, and more stirring to the emotions. The sun is the enemy whom it is urgent to avoid, to elude, or to combat. And avoiding him means, in a manner of speaking, as we have already pointed out, the inhumation of the moribund flora, the burying of its stalks in the earth. But inasmuch as the earth is in turn hard and rough, drained dry by the slopes or rendered sterile by the suction of the strata, completing the work of the sun's rays, between these two unfavorable environmental factors — the overheated atmosphere and the inhospitable soil — the most robust growths take on a highly abnormal aspect, bearing the stigma, all of them, of the silent battle that is going on.

The leguminous plants, which grow high in other places, are here dwarfed. At the same time they increase the ambit of their foliage, thus broadening the surface of contact with the air, for the absorption of those few elements that are diffused in it. Their principal roots are atrophied as a result of beating against the impenetrable subsoil, and in their place is a wide expanse of secondary radicels, clustered in sap-swollen tubercles. The leaves are increased in number and, hard as chips, are stuck rigidly on the tip of the boughs, by way of reducing the amount of surface exposed to the sun. The fruits, stiff as cones sometimes, have a protective covering, and the dehiscence with which the pods open is perfect, as if manipulated by steel springs, an admirable arrangement for the propagation of the seeds, which are scattered profusely over the ground. And they all, without a single exception, in the form of the sweetest of perfumes, possess a barrier which on cold nights is reared above them to prevent their suffering from sudden drops in temperature, charming and invisible tents to protect them. Thus does the tree apparel itself in reaction to a cruel climate.

The cautery of the droughts is adjusted on the backlands; the burning air is sterilized; the ground, parched and cleft, becomes petrified; the northeaster roars in the wilderness; and, like a lacerating haircloth, the caatinga extends over the earth its thorny branches. But the plants, all their functions now reduced, "weathering out the season," the life within them latent, feed on those reserves which they have stored up in the off seasons and contrive to ride out the ordeal, ready for a transfiguration in the glow of a coming spring.

Some of them, in a more favorable soil, owing to a most singular circumstance, are even more successful in eluding the inclemencies of their environment. Grouped in clusters or standing about isolated here and there are to be seen numerous weedy shrubs of little more than a yard in height, with thick and lustrous leaves, an exuberant and pleasing flora in the midst of the general desolation. They are the dwarf cashew-nut trees, the typical Anacardia humilis of the arid plains, the cajuys of the natives. These strange trees have roots which, when laid bare, are found to go down to a surprising depth. There is no uprooting them. The descending axis increases in size the further they are scraped, until one perceives it parting in dichotomous divisions which continue underground to meet in a single vigorous stalk down below.

These are not roots; they are boughs. And these tiny shrubs, scattered about or growing in clumps, covering at times large areas, are in reality one enormous tree that is wholly underground. Lashed by the dog-day heat, fustigated by the sun, gnawed by torrential rains, tortured by the winds, these trees would appear to have been knocked out completely in the struggle with the antagonistic elements and so have gone underground in this manner, have made themselves invisible, with only the tallest shoots of their majestic foliage showing above ground.

Others, lacking this conformation, are equipped in another fashion. The waters that flee in the savage whirl of the torrents, or between the inclined schist layers, are retained for a long time in the spathes of the Bromeliaceae, thus bringing these plants back to life. At the height of summer, a macambira stalk is for the thirsty woodsman a cup of water, crystalline and pure. The greenish thistles with their tall, gorgeous flowers, the wild pineapples, and the silk grass, growing in impenetrable thickets, all show the same form, one that is purposely adapted to these sterile regions. Their ensiform leaves, which, like those of the majority of plants in the backlands, are smooth and lustrous, facilitate the condensation of the scant vapor brought in by the winds, thus overcoming the greatest menace to vegetable life resulting from a broad field of evaporation on the leaves, which drains off and offsets the moisture absorbed by the radicels.

There are other plants equipped in a different manner, with another kind of protective apparatus, but one equally resistant. The Indian figs and cacti, native throughout the region, come under Saint-Hilaire's category of vegetable fountains. Classic types of desert flora, more resistant than the others, when whole trees blasted by lightning fall at their side, they remain unharmed and even flourish to a greater extent than they did before. Suited to cruel climates, they grow thin and etiolated in the milder ones; for the fiery environment of the desert appears to stimulate better the flow of sap in their tumid leaflike branches.

The favellas, nameless still to science — unknown to the learned, although the unlearned are well acquainted with them — a future genus cauterium of the Leguminosae, it may be, possess leaves which, consisting of cells elongated in the form of villosities, are a remarkable means of condensation, absorption, and defense. On the one hand, their surfaces, cooling at night to a degree far below the temperature of the air, despite the aridity of the atmosphere, provoke brief precipitations of dew. On the other hand, whoever touches them touches an incandescent plate heated to an unnatural degree.

When, contrary to the cases mentioned, the species are not well equipped for a victorious reaction, arrangements which are, perhaps, still more interesting may then be observed. In this case, the plants unite in an intimate embrace, being transformed into social growths. Not being able to weather it out in isolation, they discipline themselves, become gregarious and regimented. To this group belong all the Caesalpinia and the catingueiras, constituting in those places where they appear 60 per cent of the desert flora; and then there are the tableland evergreens and the pipe reeds, shrubby, hollow-stemmed heliotropes, streaked with white and with flowers that grow in spike clusters, the latter species being destined to give its name to the most legendary of villages [Canudos].

These are not to be found in Humboldt's table of Brazilian social plants, and it is possible that the first named also grow, isolated, in other climates; but here they are distinctly social. Their roots, tightly interlaced beneath the ground, constitute a net to catch the waters and the crumbling earth, and, as a result of prolonged effort, they finally form the fertile soil from which they spring, overcoming, through the capillarity of their inextricable tissues, with their numerous meshes, the insatiable suction of the strata and the sands. And they do live. "Live" is the word — for there is, as a matter of fact, a higher significance to be discerned in the passivity exhibited by this evolved form of vegetable life.

The Joaz Trees

The joaz trees show the same characteristics; they rarely lose their intensely green leaves, purposely modeled to cope with the vigorous action of the light. Scorching months and years go by, and the extremely rugged soil becomes utterly impoverished; but in those cruel seasons in which the heat of the sun's rays is aggravated by spontaneous combustions kindled by the powerful attrition of the dried and peeling branches in the high winds — even then, aliens to the march of the seasons, these trees, above life's widespread desolation all around them, have boughs that are always green and flowering and strew the desert with bright golden-hued blooms, which lie there in the gray of the stubble like festive, green-decked cases.

In certain seasons, however, the harshness of the elements increases to such a point that the joaz boughs are stripped. The bottoms of the freshwater pits have long since been buried under; the hardened beds of the ipueiras, turned into enormous molds, show the ancient print of oxen; and throughout the entire backland region life comes to a standstill.

At such times, amid this still-life scene, only the cacti stand thin and silent, their circular stocks divided into uniform polyhedral columns with the impeccable symmetry of huge candelabra. And looming larger as the short-lived afternoon comes to these desert parts, with their large and gleaming vermilion-colored fruit standing out against the half-light of dusk, they convey the moving illusion of giant tapers stuck at random in the soil, scattered over the plains, and lighted. They are characteristic of the capricious flora here in the heart of summer.

The mandacarús (Cereus jaramacarú), attaining a remarkable height, rarely appearing in groups but standing isolated above the chaotic mass of vegetation, are an attractive novelty at first. They serve as contrast, towering triumphantly while all about them the flora is depressed. One's gaze, disturbed by the painful contemplation of this maze of twisted boughs, grows tired and seeks relief in the straight and proper lines of their stocks. After a little while, however, they become an oppressive obsession, leaving on everything the imprint of an unnatural monotony, as they follow one another row on row, all uniform, identical, all of the same appearance, all equally distant, and distributed throughout the desert with a singular display of order.

The chique-chique (Cactus peruvianus) are a variant of inferior pro-portions, with curved and creeping boughs teeming with thorns and covered with snow-white flowers. They seek out the hot and sterile sites. They are the classic vegetation of parched areas. They revel in the burning bed of sun-stricken granite blocks.

They have as inseparable companions, in this habitat which even the orchids avoid, the monkshoods, inelegant and monstrous melocacti, fluted and ellipsoidal in form, with thorny buds that converge above in an apex formed by a single bright-red flower. They make their appearance in an inexplicable manner over the barren rock, really conveying, in their size, their conformation, and the manner of their dispersion, the singular impression of bloody, decapitated heads tossed here and there, without rhyme or reason, in a truly tragic disorder. An extremely narrow cleft in the rock permits them to insinuate their long capillary root until it reaches a spot down below where there may possibly exist a few drops of moisture that have not been evaporated.

The whole of this vast family, assuming all these varied aspects, gradually declines until we come to the humblest of them all in the thorny, creeping quipás, spread over the earth like a harrowing matweed; and the sinuous rhipsalides, twining like green vipers through the three branches; along with those fragile epiphytic cacti of a pale sea-green hue, clinging by their tentacles to the stalks of the urucuri palms, fleeing the unfriendly soil for the repose of their tufts.

Here and there are other varieties, among them the "devil's palms," diminutive-leafed opuntias, diabolically bristling with thorns, showing the vivid carmine of the cochineals that feed on them, and bordered with rutilant flowers; they form a bright spot that helps to break the mournful solemnity of the landscape.

There is little more to be distinguished by one who, on a clear day, strolls through these desert tracts, amid trees without either leaf or flower. All the flora, indeed, is jumbled in an indescribable and catastrophic confusion. For this is the caatanduva, the "ailing forest" in the native etymology, grievously fallen upon its terrible bed of thorns! Climb any elevation whatsoever and let your gaze wander, and it will encounter the same desolate scene: a shapeless mass of vegetation, the life drained from it, writhing in a painful spasm.

This is the silva aestu aphylla, the silva horrida of Von Martius, laying bare in the luminous bosom of tropical nature a desert vacuum. One now begins to appreciate the truth of Augustin de Saint-Hilaire's paradoxical statement: "There is here all the melancholy of winter, with a burning sun and the heat of summer!"

The crude light of the long days flames over the motionless earth without bringing it the least animation. The quartz veins on the limestone hills scattered in disorderly fashion over the desert give off a gleam of banquises; and from the dried bough tips of the numbed trees hang whitish tillandsia, like melting snowflakes, giving to the picture the aspects of a glacial landscape, with a vegetation hibernating in the icy wastes.

The Tempest

But at the end of any afternoon in March, swift-passing afternoons, without twilights, soon drowned in night, the stars may be seen for the first time, sparkling brilliantly.

Voluminous clouds wall off the far horizon, scalloping and embossing it with the imposing outlines of black-looming mountains. The clouds rise slowly and slowly puff out into great dark masses high above, while down below on the plains the winds are raging tumultuously, shaking and twisting the branches of the trees.

Clouded over in a few minutes' time the sky is shot with sudden lightning flashes, one after another, which go to deepen the impression of black tempest. Mighty thunderclaps resound, and great rains begin to fall at intervals over the earth, turning then into a veritable cloudburst.

Resurrection of the Flora

Upon returning from his stroll, the astonished traveler no longer sees the desert. Over the ground, carpeted with amaryllis, he beholds instead the triumphant resurgence of the tropical flora.

It is a complete change of scene. The rotund mulungús, on the edge of the waterpits, which are now filled, display the purpling hue of their large vermilion-colored flowers, without waiting for leaves. The tall caríbas and baraúnas, however, are leafing along the borders of the replenished streams. The stripped marizeiros rustle audibly to the passage of the fragrant breezes. The quixabeiras, with their tiny leaves and fruits like onyx stones, take on a livelier appearance. Of a deeper green, the icoseiros grow thickly over the fields, beneath the festive swaying of the urucuri tufts. The flowering thickets of tableland evergreens with their fine, flexible stocks are billowing motion, lending life to the landscape as they spread out over the plains or round out the slope of the hills. The umburanas perfume the air, filtering it through their leafy boughs. And dominating the general revival, if not by their height, at least by the beauty of their appearance, the circular-spreading umbú trees lift their numerous branches to a height of a couple of yards above the ground.

The Umbú Tree

This is the sacred tree of the backlands. Faithful companion to the cowboy's swift, happy hours and long, bitter days, it affords the most apt example of the adaptation of the hinterland flora to its environment. It may be that once it grew taller and more vigorous, and that, as the result of an alternating succession of flame-belching summers and torrential winters, it has declined to its present stature and, in the course of its evolution, has undergone modifications that give it greater resistance, enabling it to defy the prolonged droughts by sustaining itself in the seasons of misery on the copious reserves of vital energy which it has stored up in its roots in more propitious times.

This energy it shares with man. This particular section of the backlands is so sterile that there is even a scarcity of those wax palms which are so providentially distributed throughout the neighboring regions as far as Ceará; and, if it were not for the umbú tree, it would be uninhabited. The umbú is for the unfortunate woodsman who dwells here what the mauritia is to the denizens of the llanos.

It feeds him and assuages his thirst. It opens to him its friendly, caressing bosom, and its curved and interlacing boughs appear especially made for the fashioning of bamboo hammocks. And, when happy times arrive, it gives him fruit of an exquisite savor for the preparation of the traditional drink known as umbusada.

The cattle, even in times of plenty, love the acidulous juice of its leaves. It then elevates its posture, lifting the firm, rounded lines of its tuft in a perfect plane above the ground, at the height of the tallest oxen, like those ornamental plants which are commonly intrusted to the care of experienced gardeners. When plucked, these trees take on the appearance of large spherical segments. They dominate the backland flora in favorable seasons as the melancholy cacti do at the time of the summer spasm.

The Jurema

The juremas are beloved of the caboclos — their hashish, of which they are inordinately fond, supplying them gratis with an inestimable beverage that reinvigorates them after long walks, doing away with their fatigue in a few moments, like a magic potion. The juremas grow in hedges, impenetrable palisades disguised by a multitude of small leaves. There are, also, the marizeiros, mysterious trees which presage the return of rain and of the "green" season and the end of the "lean" days. When at the height of a drought, a few drops of water ooze along the withered bark of their trunks, they put forth leaves once more. The joaz trees may be bright in the copses, and the baraúnas with their clustering flowers, and the araticús on the edge of the brooks. But, nonetheless, the umbús, adorned with their snow-white blossoms and budding leaves, of an elusive shade that ranges from a pale green to the vivid rose of the newest shoots, stand out as the brightest spot in the whole of this dazzling scene.

The Backlands are a Paradise

The backlands are now a paradise.

At the same time the hardy fauna of the caatingas comes to life. From the damp lowlands the caitetús start. Ruddy-legged boars go through the plantations in droves, with a loud crunching of jaws. Herds of swift rheas run over the high tablelands, spurring themselves on with their wings. The mournful-voiced seriemas and the vibrant-toned sericoias sing in the undergrowth along the brooksides, where the tapir comes to drink, pausing for a moment in his ugly canter, which is in an unvarying straight line across the caatingas, to the destruction of trees in his path. Even the pumas, frightening the timid mocos which seek shelter in their rock burrows, skip merrily about in the tall weeds before settling down in ambush for the coy stag or straying bullock.

Hinterland Mornings

There come then matchless mornings, as the radiating glow from the east once more tinges the erythrinas with purple and wreathes with violet the bark of the umburanas, setting off to better advantage the multicolored festoons of the bignonias. The air is animated by the palpitation of swift, rustling wings. The notes of strange trumpets strike our ears. In a tumult of flight, this way and that, flocks of beautiful homeward-bound pigeons pass overhead, and one can hear the noisy, turbulent throngs of maritatacas. Meanwhile, the happy countryman, forgetful of former woes, makes his way down the trails, driving a full-bellied herd and humming his favorite air.

And so the days go by. One, two, six fortunate months, blessed by the earth's abundance, elapse, until silently, imperceptibly, with an accursed rhythm, the leaves and flowers gradually begin to fall, and the drought once more descends on the dead boughs of the shorn trees.

The Lagoon*

James Morris

Sometimes in a brutal winter night you may hear the distant roar of the Adriatic, pounding against the foreshore; and as you huddle beneath your bedclothes it may strike you suddenly how lonely a city Venice remains, how isolated among her waters, how forbiddingly surrounded by mud-banks, shallows and unfrequented reedy places. She is no longer a true island, and the comfortable mainland is only a couple of miles from your back door; but she still stands alone among the seaweed, as she did when the first Byzantine envoys wondered at her gimcrack settlements, fourteen centuries ago. Time and again in Venice you will glance along some narrow slatternly canal, down a canyon of cramped houses, or through the pillars of a grey arcade, and see before you beneath a bridge a tossing green square of open water: it is the lagoon, which stands at the end of every Venetian thoroughfare like a slab of queer wet countryside.

Several sheltered spaces of water, part sea, part lake, part estuary, line the north-western shores of the Adriatic: in one Aquileia was built, in another Ravenna, in a third Comachhio, in a fourth Venice herself. They were known to the ancients as the Seven Seas, and they were created in the first place by the slow action of rivers. Into this cranny of the Mediterranean flows the River Po, most generous of rivers, which rises on the borders of France, marches across the breadth of Italy, and enters the sea in a web of rivulets and marshes. Other famous streams tumble down from the Alpine escarpment, losing pace and fury as they come, until at last they sprawl sluggishly towards the sea in wide stone beds: the Brenta, which rolls elegantly through Padua out of the Tyrol; the Piave, which rises on the borders of Austria, and meanders down through Cadore and the delectable Belluno country; the Sile, which is the river of Treviso; the Adige, which is the river of Verona; the Ticino, the Oglio, the Adda, the Mincio, the Livenza; the Isonzo; and the Tagliamento. This congregation of waters, sliding towards the sea, has made the coastline a series of estuaries, interlinked or overlapping; and three rivers in particular, the Piave, the Brenta and the Sile, created the Venetian lagoon. If you look very hard to the north, to the high

Alpine valleys in the far distance, lost among the ridges and snow peaks, then you will be looking towards the ultimate origins of Venice.

When a river pours out of a mountain, or crosses its own alluvial plain, it brings with it an unseen cargo of rubble: sand, mud, silt, stones and all the miscellaneous bric-à-brac of nature, from broken tree-trunks to the infinitesimal shells of water-creatures. If the geological conditions are right, when its water eventually meets the sea, some of this material, buffeted between fresh water flowing one way and salt water pushing the other, gives up the struggle and settles on the bottom, forming a bar. The river forces its way past these exhausted sediments, the sea swirls around them, more silt is added to them, and presently they become islands of the estuary, such as litter the delta of the Nile, and lie sun-baked and turtle-haunted around that other Venice, the southern-most village of the Mississippi.

Such barriers were erected, aeons ago, by the Brenta, the Piave and the Sile, when they met the currents of the Adriatic (which, as it happens, sweep in a circular motion around this northern gulf). They were long lonely strips of sand and gravel, which presently sprouted grass, sea-anemones and pine trees, and became proper islands. Behind them, over the centuries, a great pond settled, chequered with currents and counter-currents, a mixture of salt and fresh, an equilibrium of floods; and among the water other islands appeared, either high ground that had not been swamped, or accumulations of silt. This damp expanse, speckled with islets, clogged with mud-banks and half-drowned fields, protected from the sea by its narrow strands — this place of beautiful desolation is the Venetian lagoon. It is thirty-five miles long and never more than seven miles wide, and it covers an area, so the most confident experts decree, of 210 square miles. It is roughly crescent-shaped, and forms the rounded north-western corner of the Adriatic, where Italy swings eastward towards Trieste and Yugoslavia. Its peers among the Seven Seas have long since lost their eminence — the lagoon of Ravenna silted up, the lagoon of Aquileia forgotten: but the lagoon of Venice grows livelier every year.

Very early in their history, soon after they had settled on their islands and established their infant State, the Venetians began to improve upon their bleak environment. It was a precarious refuge for them. The sea was always threatening to break in, especially when they had weakened the barrier islands by chopping down the pine forests. The silt was always threatening to clog the entire lagoon, turning it into a vulnerable stretch of land. The Venetians therefore buttressed their mud-banks, first with palisades of wood and rubble, later with tremendous stone walls: and more fundamentally, they deliberately altered the geography of the lagoon. Until historical times five openings between the bars — now called lidi — connected it with the open sea, allowing the river water to leave, and the Adriatic tides to ebb and flow inside. The Venetians eliminated some of these gaps, leaving only three entrances or porti through which the various waters could leave or enter. This strengthened the line of the lidi, deepened the remaining breaches, and increased the scouring force of the tide.

They also, in a series of tremendous engineering works, diverted the Brenta, the Sile, the Piave and the most northerly stream of the Po, driving them through canals outside the confines of the lagoon, and allowing only a trickle of the Brenta to continue its normal flow. The lagoon became predominantly salt water, greatly reducing (so the contemporary savants thought) the ever-present menace of malaria. The entry of silt with the rivers was virtually stopped: and this was opportune, for already half the lagoon townships were congealed in mud, and some had been entirely obliterated.

Thus the lagoon is partly an artificial phenomenon; but although it often looks colourless and monotonous, a doleful mud-infested mere, it is rich in all kinds of marine life. Its infusions of salt and fresh water breed organisms luxuriantly, so that the bottoms of boats are quickly fouled with tiny weeds and limpets, and the underneaths of palaces sprout water-foliage. The lagoon is also remarkable for its biological variety. Each porto governs its own small junction of rivulets, with its own watershed: and wherever the tides meet, flowing through their respective entrances, there is a recognizable bump in the floor of the lagoon, dividing it into three distinct regions. It is also split into two parts, traditionally called the Dead and the Live Lagoon, by the limit of the tides. In all these separate sections the fauna and flora vary, making this a kind of Kew Gardens among waters; it used to be said that even the colour of the currents varied, ranging from yellow in the north by way of azure, red and green to purple in the extreme south.

In the seaward part, where the tides run powerfully and the water is almost entirely salt, all the sea-things live and flourish, the mud-banks are bare and glutinous and the channels rich in Adriatic fish. Farther from the sea, or tucked away from its flow, other organisms thrive: beings of the marshes, sealavenders, grasses and tamarisks, swamp-creatures in semistagnant pools, duck and other birds of the reeds. There are innumerable oysters in these waters, and crustaceans of many and obscure varieties, from the sea-locust to the thumb-nail shrimp; and sometimes a poor flying-fish, leaping in exaltation across the surf, enters the lagoon in error and is trapped, like a spent sunbeam, in some muddy recess among the fens.

A special race of men, too, has been evolved to live in this place: descended partly from the pre-Venetian fishing communities, and partly from Venetians who lingered in the wastes when the centre of national momentum had moved to the Rialto. They are the fittest who have survived, for this has often been a sick lagoon, plagued with malaria, thick with unwholesome vapours, periodically swept by epidemics of cholera and eastern disease. Like the rest of the fauna, the people vary greatly from part to part, according to their way of life, their past, their degree of sophistication, their parochial environment. Inshore they are marsh-people, who tend salt-pans, fish among grasses, and do some peripheral agriculture. Farther out they can still be farmers or horticulturists, if they live in the right kind of island; but they are more likely to be salt-water fishermen, either taking their big boats to sea, or hunting crabs, molluscs and sardines among the mud-banks of the outer lagoon.

Their dialect varies, from island to island. Their manners instantly reflect their background, harsh or gentle. They even look different, the men of Burano (for instance) tousled and knobby, the men of Chioggia traditionally Giorgio-nesque. The lagoon islands were much more independent in the days before steam and motors, with their own thriving local governments, their own proud piazzas, their own marble columns and lions of St. Mark: and each retains some of its old pride still, and is distinctly annoyed if you confuse it with any neighbouring islet. 'Burano!' the man from Murano will exclaim. 'It's an island of savages!' — but only two miles of shallow water separates the one from the other.

The lagoon is never complacent. Not only do the tides scour it twice a day, the ships navigate it, the winds sweep it coldly and the speed-boats of the Venetian playboys scud across its surface in clouds of showy spray: it also needs incessant engineering, to keep its bulwarks from collapsing or its channels silting up. The Magistracy of the Waters is never idle in the lagoon. Its surveyors, engineers and watermen are always on the watch, perennially patching sea-walls and replacing palisades. Its dredgers clank the months away in the big shipping channels, looming through the morning mist like aged and arthritic elephants. The survival of Venice depends upon two contradictory precautions, forming themselves an allegory of the lagoon: one keeps the sea out; the other, the land. If the barrier of the outer islands were broken, Venice would be drowned. If the lagoon were silted up, her canals would be damned with mud and ooze, her port would die, her drains would fester and stink from Trieste to Turin (it is no accident that the romantic fatalists, forseeing a variety of dramatic ends for the Serenissima, have never had the heart to suggest this one).

So when you hear that beating of the surf, whipped-up by the edges of a bora, go to sleep again by all means, but remember that Venice still lives like a diver in his suit, dependent upon the man with the pump above, and pressed all about, from goggles to lead-weighted boots, by the jealous swirl of the waters.

On the Persistence
of Local Populations

A local population, whether it consists of men or grasshoppers, is one that persists generation after generation. Each generation gives rise to the next by the mating of parental individuals; as the younger generations mature, the older ones age and disappear. The local population that I describe in this essay is a population of cross-fertilizing organisms; in this sense, bacteria that reproduce asexually or even such plants as beans that pollinate themselves do not form local populations. Nor is a local population the population of the census taker. Our local population is one with a time dimension; it persists through successive generations.

The purpose of this essay is to examine precisely what "the persistence of a local population" means. Under what conditions do populations actually persist? How can the great fluctuations in population size be incorporated into the notion of persistence? Under what conditions do populations cease to exist? The monkeys imported to Las Marismas from Gibraltar disappeared and the camels, although they have been in southern Spain for nearly two centuries are about to disappear. Why? Why can Michener say that there will be no "wild" camels in Spain within a decade or so?

A population persists through time only if one generation appears (hatches, sprouts, is born, or whatever the appropriate verb may be) to take the place of the one that has vanished. In the case of annual plants such as marigolds, the new generation literally substitutes for an old one that has disappeared; in populations with overlapping generations, such as populations of men, the new generation physically replaces a mixture of old and fairly old individuals who have died in recent times. Whether the organism is an annual or not, however, the population persists only if, in the long run, each mother leaves an adult, fertile daughter as her replacement. Were she to leave more than one daughter, the population would expand to fantastic numbers in a remarkably short time. Darwin calculated that a pair of elephants, one of the slowest breeding of all animals, would leave 15 million descendants in 500 years if allowed to breed at the maximum rate possible. Since the number of elephants does not increase 7.5-millionfold every five centuries, their effective rate of reproduction must be much nearer the one mother-one adult daughter mentioned above than the maximum imaginable for elephants.

If populations really persisted by means of the one-for-one replacement described here, they would lead a precarious existence indeed. First of all, they would never expand in numbers. Second, the most trivial, novel challenge would most likely exterminate them. Cats, dogs, woodchucks, and other small mammals that fall victim to the automobile – not to mention the innumerable insect species whose members are splattered on windshields and embedded in automobile radiators – would by now be well on their way to extinction. The automobile is obviously not exterminating all of these organisms; so we must conclude that successive generations need not replace one another by means of carefully measured, identical numbers of individuals.

Populations do indeed grow in size with passing generations, in many instances only to shrink again. Such cyclic changes in population size are most obvious in the case of summer insects for which a generation may take two weeks or less. Persons who keep fresh fruit on their dining room tables can attest to the increase in numbers of the small vinegar fly (*Drosophila melanogaster*). Sometime in May one or two of these flies may be seen hovering about a bunch of bananas; a similar bowl of fruit in July or August may attract 50 or more flies. The fruit counter at the local grocery store may have tremendous numbers of these harmless insects unless the owner keeps a scrupulously clean establishment or (a practice that should be discouraged) uses insecticides liberally.

Over short periods of time (weeks or months for a rapidly reproducing insect, decades for a large mammal or tree) the mothers of one generation may leave more than one adult daughter and so the population grows in size as a consequence. If the number of daughters per mother is five, the population size in successive generations increases 1 to 5 to 25 to 125 and so forth. For short periods of time, populations of small insects may expand 50 to 100-fold per generation (1 to 50 to 2500 to 125,000 or 1 to 100 to 10,000 to 1,000,000). Populations of large animals (including man) do not expand at such tremendous rates because females cannot bear young in such numbers; nevertheless, as we saw earlier in discussing numbers of persons, populations that double in size every generation can reach enormous sizes in surprisingly few generations.

The long-range persistence of populations, therefore, depends upon an *average* number of daughters per mother not exceeding one. If the persistence is to be more than an accidental persistence however, this average must be attained by regulatory forces that lead to *one* as the mean number over extended periods of time.

One set of regulatory forces is effective in halting the expansion of populations; this set may involve space, food, or the growth of predator populations. It may involve the maintenance of a relatively secure segment of the population that is entrenched in suitable nesting sites or food-gathering territories and the exposure of the remainder to more dangerous circumstances.

In this case, for example, the raccoons or opossums that are killed by automobiles may be regarded as wanderers; if they had not been killed on the highway, they would most likely have been eaten by foxes, killed by owls, or would otherwise have died without reproducing. Whatever the precise nature of the limiting mechanism, we can see intuitively that upper limits on the number of plants or animals of any sort must exist because the earth is finite in size; furthermore, we can see that this limit is reached rather quickly by any population that increases in an exponential fashion.

There is another side to the story of regulation, a second set of regulatory forces. Small populations tend to expand because females of most species can produce many young. After a population has reached its peak expansion, it contracts. Occasionally contracting populations vanish altogether, but more likely they persist at a low level only to expand again when favorable times (late spring and summer for the vinegar fly) come again. In this case, population regulation is such that the few surviving individuals have an excellent chance of leaving numerous offspring.

For any particular local population such as that of the native, dark-bodied fruit fly (*Drosophila athabasca*) living in my back yard, all individuals may disappear at any time – during a severe winter for example. In this case, the next population of these flies will be descendants of chance migrants. On the one hand, local populations are exposed to constraints that act in opposition to unlimited population growth. On the other hand, as the numbers of individuals dwindle, they are relieved of subsidiary pressures; hence, each has an increasing probability of surviving and of leaving surviving offspring. These are the opposing regulatory mechanisms that prolong, even though they may not guarantee, the persistence of local populations.

Boom and Bust:
The Biology of Calamities

The preceding essay dwelt on the persistence of local populations, on the occurrence of a given species of plant or animal in a given area year after year after year. Persistence unaccompanied by gross systematic changes in population size suggests that each female of one generation leaves almost precisely one adult female replacement in the succeeding one (males are ignored in this discussion as a matter of convenience; if they were to be included, we would claim that each breeding pair leaves an average of two adult offspring as their replacements). In discussing the persistence of populations, however, I acknowledged their periodic expansions as well as subsequent contractions. These expansions and contractions can involve tremendous numbers of individuals. Consider, for example, the small vinegar fly (this fly is a recurrent subject in these essays because it is my favorite research material) in a vineyard in upstate New York. At the height of the grape season when ripe fruit can be found for miles on end and when tons of waste pressings accumulate near the wineries, millions and millions of these flies exist. Following the first killing frost, however, and during the subzero winter, only an occasional fly survives — in a basement or perhaps a vegetable cellar — to start the cycle once more when spring arrives.

In this essay I shall consider the expansion and collapse of populations as "boom and bust" and concentrate on such cycles in relation to still other populations. I shall, in short, consider how the dying carp of Las Marismas influence the populations of hawks and kites, or how the large numbers of bewildered ducks searching for vanishing water holes fall prey to the fox and the lynx. The catastrophe that causes the crash of one species very often brings on the explosive growth of another; such is the biology of catastrophe.

Natural catastrophes may, for our present purposes, be classified as sporadic and major, cyclic and major, or sporadic and minor. A forest fire, for example, would qualify as a major, sporadic catastrophe; the annual destruction of ducks and carp in Las Marismas or of fruit flies at the onset of winter is a cyclic catastrophe; the occasional loss of life in the quicksand of the Marismas or in the La Brea tar pits of Los Angeles is a sporadic, minor catastrophe — an individual catastrophe.

Major catastrophes, even though sporadic, can provide a way of life for many types of animals. Ivory-billed woodpeckers, once throve on the rotting, insect-infested timber of burned-over forests; the prevention of forest fires and the general reduction of forested areas have so restricted the livelihood of these birds that they now appear to be extinct. Moose, too, depend upon forest fires to provide them with a great deal of food because they browse on the small elders that spring up in recently burned-over meadows and woodlands. As long as a major catastrophe such as a forest fire occurs frequently enough within the roaming area of a population of woodpeckers or moose, these catastrophes can be relied upon and utilized as a means of livelihood. Once such catastrophes become rare, both in the sense of infrequent in time and widely dispersed in space, then they no longer afford a reliable way of life for dependent species, and these species dwindle and disappear.

Individual catastrophes – accidental deaths of various animals – are the means of support for vultures, hyenas, and all such carnivores that do not generally catch their own prey. These scavengers are served by standing and waiting. Once more, however, the supply of food must be reliable enough to offer continuing support for the dependent species. A high-speed thruway can provide an unexpected bonanza for the buzzards of our Southeastern states; a strategically placed watering trough on a Western cattle ranch may doom many of the vultures of that area.

The cyclic booms and busts that occur at annual intervals are perhaps the most interesting of all catastrophes because of the solvable problems they pose for other organisms and the interlocking links in the overall chain of life that develop as a consequence. A rapidly breeding insect may swell in numbers hand-in-hand with the development of a suitable source of food; aphids, for instance, can increase in number directly as the availability of rose leaves or other succulent greenery increases. The same sequential arrangement does not hold in the case of larger animals that depend upon the presence of thriving populations of small ones. Gnatcatchers, insectivorous birds, may feed their nestlings at the rate of nearly one feeding per minute. To sustain this rate requires an already thriving population of small insects of various sorts. The gnatcatchers, however, lay their eggs and incubate them in anticipation of the swelling insect population. The mechanisms upon which such anticipation is based are fascinating and will be dealt with in the following essay; for the moment, the interesting point is the utilization of a boom-time population of insects. The birds remove insects from a thriving population that would be removed (or whose immediate descendants would be removed) by the onset of drought or cold weather.

Gnatcatchers reap the thriving insect populations before they would otherwise disappear. The kites and hawks of Africa reap the carp in Spain as the streams dry up in the summer. The lynx and the fox catch the ducks that are left without water holes necessary for their survival. Predation, therefore, is a form of harvesting – the harvesting of a surplus population that is destined for destruction under any circumstances. For the predator, getting food is easiest at the height of the harvest, and it is at this time that the predator generally chooses to produce his own young. And so the boom in the prey population becomes altered into a boom for the predator; the "bust" of the prey population is the next step.

I once explained to a young child about flies whose eggs are discovered and parasitized by a small wasp. I asked her what would happen if the fly laid more eggs. Would she expect larger number of flies? Or a larger number of wasps? She answered, "Wasps," the correct answer under at least a large number of conditions. Here was an example of excellent ecological insight: fly eggs make wasps, not flies; carp make hawks and insects make gnatcatchers. The course of food through the web of life is never ending.

Events, Random and Cyclic: Problems of Anticipation

The previous essay touched briefly on the matter of anticipation; this essay deals with this fascinating question in greater detail. And anticipation is fascinating indeed. We saw, for instance, that gnatcatchers must have access to enormous numbers of small insects in order to feed their fledglings. The population of suitable insects swells in numbers abruptly, persists for a short time, and then collapses once more. The production of a brood of fledglings, on the other hand, follows from an extended series of physiological events, a series that must be initiated long before the insect population explodes. The gnatcatchers, in a sense, place their future on the line when they start their reproductive cycle; they gamble that fledglings to be fed and abundant food to be caught will be coincident events. How is this gamble made? If we understand this example, we will understand why physiologists and ecologists (or naturalists, if the ecologists will pardon this term) differ in assigning causes to observed events. We shall also see why the harsh, rhythmic annual cycles of Las Marismas of Spain or of the Caatingas of Brazil are simpler challenges to the inhabitants of these areas than are random fluctuations that are seemingly much less severe. We shall see how fortune-telling and reading the future are built into the genetic makeup of a species.

A pair of gnatcatchers do not simply meet, mate, lay eggs, and produce offspring. The reproductive cycle of any organism, but especially of higher organisms, is much more complex. Limiting ourselves just to the maturation of the testes and ovaries, we find that these reproductive organs in a bird are under the control of hormones secreted by various glands such as the pituitary in the brain and the thyroid. These and other glands indulge in an interplay of stimulation and repression that leads to sexual maturity, nest building, and mating activity. The female lays several eggs; these are incubated, and finally they hatch. Overall, a considerable amount of time may elapse while the entire sequence of events runs its course.

The growth of small-insect populations, in the meantime, is following its own course. Some plants are sending out leaves; the fruit of others may have ripened and have fallen. Short-lived insects with rapid reproductive rates are exploiting these sources of food as they become available and are increasing tremendously in number, 10-fold or 100-fold each generation. Very little or no anticipation is required on the part of the insects; they simply exploit available material and exploit it instantaneously.

The key to the development of the insect population is to be found, of course, in the expansion in the amount of plant material that serves as food. Now, plants do not sprout, or send out leaves, or even shed pollen in a willy-nilly fashion; they read the environment rather carefully while undergoing developmental processes. Lilac leaves, for example, expand gradually throughout the spring; their size serves as an extremely accurate indicator of the cumulative degree-days of the preceding days and weeks. Many plants respond to photoperiod, the time between sunrise and sunset; they put out leaves or bloom only when daylight has reached a certain number of hours. Because days become longer in the spring, hours of daylight are a fair indicator of the temperature; after the day has attained a certain length, killing frosts are unlikely.

The trigger for plants turns out to be the trigger for many birds as well; that is the hormonal activity in birds is known in many instances to be stimulated by the number of daylight hours. In a laboratory birds can be brought into sexual activity at odd times of the year simply by regulating the timing of the lighting system.

The manipulation of sexual activity in birds by the manipulation of the laboratory lighting system leads the physiologist to say that the bird's hormonal system is under the control of light, that it is photoperiodic. This is the immediate or proximal control of the reproductive cycle, to be sure. The ultimate cause, however, is quite different. Photoperiod offers the bird a clue as to the coming time at which food will be plentiful; it sets his reproductive machinery in gear and leads to the synchronization of mouths to feed and food to be had. The precise adjustment that permits the machinery to start just at the time needed to make the two events actually coincide is brought about by natural selection. Nothing succeeds like success and, generation after generation, those individuals with the more nearly perfect timing are those that leave the most surviving offspring.

The test for some of the preceding claims lies in the exceptions. The initial stimulant for plants in a desert is not length of day but rain; at the first substantial rainfall of the rainy season, seeds germinate, plants shoot up, flowers bloom, but in a short time the desert reverts to its parched self once more. Desert birds are stimulated to enter their reproductive cycle by rain; the number of daylight hours is not effective for these birds. Birds that live on seeds and grain need not be stimulated by light in the same manner as are insect feeders. The wood pigeon of Britain, for example, starts its reproductive cycle but the female's ovaries enter a "hold" position (just as in the countdown during a rocket launching) from which they are reactivated only by grain and seeds. The sexual maturity of those birds, consequently, coincides with the timing of the local harvest. The ultimate cause for the reproductive pattern in each of these cases clearly is the production of young at the very moment when the available food is at its highest level.

The tuning-in of an animal or plant species to the cyclic changes in the world around it requires the use of some aspect of the cycle as a signal: long hours of daylight mean warm weather; rain in a desert means plants will appear. As long as a signal exists to be read, the severity of the intervening periods is relatively unimportant. Adaptations that permit an organism to cope with severe conditions can generally be found; under the influence of natural selection, plants can develop and have developed mechanisms for resisting high or low temperatures as well as wet or dry surroundings. Insects can overwinter in temperatures much below freezing. Dormant seeds of most plants are virtually inert. The severity of conditions during dormant periods is not a serious problem; the vital condition, instead, is the presence of a signal that can be used to anticipate coming events and that is in fact used to initiate changes that render the individual vulnerable in case the predicted or anticipated events fail to materialize.

Random events, by definition, cannot be anticipated. They are not preceded by a reliable signal that can be used to initiate a series of predictive actions. They have the effect of rendering the adaptation of a species to its environment less than perfect. The lack of perfection manifests itself in variation within the species. Some birds enter sexual activity before others; the nesting times of birds of the same species differ slightly; some females lay more eggs than others; and incubation times differ slightly. As a group these variations represent a lack of perfection in the adaptation of the species; over extended periods of time, however, they allow the species to alter responses to the triggering signal whenever an alteration becomes advantageous.

The difficulties inherent in any attempt to adjust to random events are not easily illustrated by the failure of adaptive changes in populations. They can best be illustrated by citing the random flight patterns that are characteristic of many startled insects. These insects, if disturbed while at rest, take flight in an entirely erratic and randomized pattern. Confronted with such a flight pattern, a predator is unable to calculate an anticipated point at which the fleeing insect can be intercepted. Were the escape flight based on any rational or stereotyped system, the predator in theory could develop an analyzing device that would predict a point of capture. Random flight offers no such opportunity. And what is true of flight is true of longer-spaced climatic events and the adaptation of a species to its environment; events that come upon the species in an unpredictable manner cannot be anticipated and, consequently, cannot be met by physiological responses that require extended periods of time to run their course following their initiation.

Predation: Harvesting the Population Surplus

Once or twice each month during the summer my wife tearfully announces that our cat has caught another bird and that the feathers must be cleaned from the back porch. The cat and I take this news much better than my wife or the bird – the cat, because he could not care less; I, because I realize that excess birds must disappear somewhere and why not on my back porch. My wife, on the contrary, suspects that each bird removed from the neighborhood is one less bird in the universe and, hence, that all birds will eventually disappear if cats have their vicious feline ways. This essay is written to convince my wife and others of her persuasion that predation is normally a form of harvest and, in fact, that a population that goes unharvested is at times in serious trouble.

On a number of occasions I have claimed that populations can expand in size – in number of individuals – at tremendous rates if left unmolested. The tendency to expand lies in the large number of offspring produced by each pair of parents – several hundred for a pair of flies, five or six for a pair of robins or starlings, three or four or sometimes more per human couple. The increase in size of an uncontrolled population is exponential; the larger the population, the larger the new increment added by a single generation.

In a stable population the excess production must be removed. It must be removed either gradually on a generation-by-generation basis or abruptly at the termination of a boom period when the population "crashes" to its normal size. The excess, to repeat, *must* be removed; there is no alternative in a world of finite size. With this matter settled, the question arises, "*Who will be removed?*"

The harvested individuals of any prey population are not a random sample of the individuals of that population. Very frequently they are individuals who were unable to stake out and hold onto a territory of their own. Yearling birds of many species return to their birthplace and compete with older birds for nesting sites and home territories; some manage to displace previous owners, many do not. The birds, old and young, that are without territories are harried by others and become the victims of predators more often than do the legitimate home owners.

Predators are rather skillful in estimating the cost of the meal they are contemplating; housewives are not the only shrewd shoppers. Wolves may test out 20 herd of caribou before picking an animal for attack. Hawks test out small birds with feints. Mink may ignore muskrats, their favorite food animal, even while actively prowling for a meal. It seems as if predators seek the crippled, the young, and the panic-stricken; indifference on the part of an otherwise legitimate victim appears to signal to the predator that his efforts will not be amply rewarded, that a particular chase and struggle simply would not be worth the effort. This conclusion repeats a point made in an earlier essay entitled "On the high cost of food."

Man is a great harvester of wildlife. At times his procedures are rational; more often they are disastrous. At the present time, however, I believe that most large game animals within the United States are protected by law and that the regulations governing the hunting of these animals are based on the control of population size and the continued assurance of healthy game animals. Deer, I know, are in constant danger of overgrazing their ranges if not controlled by predation or by hunters; in many states bucks can be taken every year while does may be assigned special seasons of their own if more stringent population control is required.

The collection of tern and coot eggs by the Spaniards in Las Marismas represents an annual harvest of hundreds of thousands of eggs that has been taking place for centuries. Is this not a dangerous threat to the bird life of these marshes? Apparently not. The Dutch ornithologist, H.N. Kluyver, has studied the effect of removing eggs from the nests of the great tit on an isolated island in the North Sea. For five years he made a census of breeding birds in the woodland of this island; he found an average of 50 banded (his bands) and 20 unbanded (immigrants? overlooked?) birds per year. Beginning in the fifth year he destroyed about 60 per cent of the fledglings each year for four years. During these years the average number of breeding birds on the island was 55 banded and 7 unbanded. The number of breeding birds remained virtually constant. The average number of fledglings per pair on this island was 11; the destruction of 6 or 7 of these by Dr. Kluyver had no effect on the size of the population of great tits. The birds removed deliberately by Dr. Kluyver would have been removed in other ways under normal circumstances.

Not all harvests are made as sensibly as those of the coot and tern eggs by the Spaniards. The passenger pigeon was hounded to death by hunters in the United States during the late 19th and early 20th century. Individual flocks of these birds at one time contained as many as one or two *billion* individuals. Forests that served as nesting areas — hundreds of square miles — were over-whelmed by the weight of roosting birds. The flocks frequently traveled hundreds of miles each day from nesting to feeding areas only to return to the nesting area in the the evening. It appears that the strategy for survival used by these birds before the arrival of white settlers was to shift their nesting areas periodically; in this way they removed themselves from the predators that thrived on the quantity of pigeon these enormous flocks presented. When the populations of predators swelled in number (boom time for the predator), the prey population simply and suddenly vanished to another area. Man destroyed this strategy. He appeared at *all* nesting sites. Special trains brought hunters to nesting sites as these were discovered. The result of man's uncontrolled slaughter of these birds, man's blind harvest, was that the last passenger pigeon died in a Cincinnati zoo in 1914.

Life must reproduce itself; populations that appear infinite in size can disappear quickly if the individual members cannot reproduce. Uncontrolled slaughter at the time and place of breeding can eliminate any species. On the other hand, if the individuals are permitted to reproduce, an astonishingly large part of the population can be harvested by man or by natural predators without endangering the continued existence of the population itself. Reproduction is a multiplicative process that at some point must be restrained. Wildlife conservation is the art of setting reasonable times for and amounts of harvest for various game animals.

On the Maintenance of Diversity in Natural Communities

Despite the sarcastic and not-so-subtle comments of the family and neighbors, my lawn is not a natural community. Nevertheless, upon glancing down from my writing, I can see at least a dozen plant species: grass, clover, chickweed, dandelion, plantain, peppermint, sour grass, moss, chickory, and (off to the side a bit) golden rod, buttercup, violet, wild carrot, and burdock. These same plant species have been here nearly as long as I have — 10 years or more. The chickweed and the dandelions have spread somewhat and the grass has receded but otherwise there has been no striking change. The composition of my lawn seems to be, if not in equilibrium, at least quasi-stable.

A meadow, of course, is immensely richer in the diversity of plant life than is my lawn. A competent botanist can easily spot 50 or more plant species in a small patch of meadow. The total diversity of any community is much larger than an estimate based on rather large plants alone; the insects, small mites and spiders, and microscopic soil organisms are all members of a small meadow community. Larger animals such as amphibia, snakes, birds, and mammals must be included in the enumeration of life forms in the communities of still larger areas.

In this essay I shall touch on the origin of the diversity of life in a natural community but I shall not go deeply into this matter. What I shall do is to consider some of the keys to the maintenance of existing diversity. This will lead me to comment on the gross errors to which a casual evaluation of biological communities might lead in a man-manipulated environment.

The diverse forms of life that make up a biological community are found together because, on the one hand, they do not get in one another's way and, on the other, they rely upon one another in some manner. The simplest form of reliance is that of a "food chain." Plants grow because soil, water, and sunlight are available. Herbivores such as aphids, caterpillars, beetles, or certain animals live on the plants. Carnivores, including insectivores, live on the herbivores. Larger carnivores live on the smaller ones. And finally the decomposers, the bacteria and fungi, live on the carcasses and droppings of the rest of the community. The relative proportions of each of the succeeding levels in the chain are fixed within rather narrow limits because each manages to extract only about 10 per cent of the energy accumulated by the lower one. Of 1,000 calories stored in plants, butterfly larvae may collect 100, of which 10 are obtained by neighborhood birds, and 1 goes eventually to my cat and his friends. A food chain cannot be exceedingly long; five or six steps seem to be the maximum theoretically possible.

Food chains are never simple linear structures; they interlock by means of cross-linkages. Our cat catches not only birds but also mammals – herbivores and carnivores; many animals eat both plants and animals. The "food chain," therefore, is not really a chain but rather a web or a coat of mail. The greater the complexity of the food web, the more stable it proves to be in the face of small calamities. The flow of energy from one level to another in a complex web has many routes, and so the community as a whole has many alternate means by which its equilibrium composition can be maintained. Similarly, the modern industrial conglomerate is a community of industries that is formed ostensibly to exploit the stability that diversification brings.

In addition to the dependence of one member of the community on the next, members of communities can live together in part because they do not interfere with one another. This can be regarded as a "packing" effect, the fitting together of many species and the consequent efficient utilization of local resources. Grasses might send out roots in the upper surface of the soil, consequently, the lower depths are available for large plants or plants with tap roots. Deep groundwater may be available to some plants, surface water to others, and dew might be used by still others. This type of sharing of resources will serve to start any primitive community. The mere coexistence of plants or animals, however, is sufficient to initiate additional selective processes that lead to the dependence of one type on another – the dependence of the morning glory upon a tree trunk for support for example. The selective processes set in motion by coexistence lead to the mutual coadaptations exemplifying communal life.

Communities represent stable associations of plants and animals; that is, the composition will return to its original state if disturbed or alternatively, will change very slowly like the succession of forms in an abandoned field. Occasionally, the apparent stability is largely dependent upon a key species. Should the key be removed, the community disintegrates. Fine examples of keys, and of the subtleties involved in identifying them, are to be found in the meadow communities of small islands off the coast of Britain.

A number of these small coastal islands have been surveyed periodically by British botanists for two centuries or more. The number of plant species growing in the meadows is frequently as high as 60 or 70. Now, these meadows are the home for the European rabbit; the rabbits are the herbivores that keep the meadows cropped.

The European rabbit is extremely susceptible to a disease, myxomatosis, that is caused by a smallpox-like virus. When a newly exposed rabbit population is infected, the mortality rate may be 98 per cent or more. The myxoma virus was used in an effort to eradicate the hordes of European rabbits in Australia; it was extremely effective. Either by accident or design, these viruses have been introduced at several places in Europe.

The rabbit population of one of those small, offshore islands was destroyed by the introduction of myxoma virus. The immediate effect on the local plant community was a sudden increase in the total number of species; perennial species whose existence on the island had never been detected during centuries of botanical surveys suddenly came to light. The burst in the number of species was short-lived however; within two or three years grasses had overrun the meadows, and the total number of plant species on the island had dropped to about five. The former richness of the island's meadow community, consequently, can be ascribed to the key species, the rabbit. It was the grazing of the rabbit that kept down the grasses and permitted the wealth of other species to thrive.

The identification of the *key* species that is largely responsible for the observed diversity is not as simple as we have just suggested. The rabbit has been identified as the key to the island meadow community. But, what about the myxoma virus? It was, after all, the presence or absence of the myxoma virus that determined in our example whether the plant community would consist of 60 to 70 species or five. But even this added possibility does not exhaust the list. On one island the myxoma virus has been introduced but there are no rabbit fleas on that island to act as vectors for the disease (in Australia the mosquito is the insect vector that carries the infection from one animal to the next). Hence, the expected epidemic has never occurred. In this case, one may claim that the rabbit flea is the key to the diversity of the meadow community; had the flea been present on the island, the number of plant species would have plummeted.

The logical difficulties that arise in attempting to identify a key species in an interlocking ecological system are identical to those faced by early geneticists as they discovered mutant genes that altered the appearance of experimental organisms. In the fruit fly, the mutation of a gene at a particular locus on the chromosome might lead to white instead of red eyes on affected individuals. This observation led some early geneticists to suggest that the nonmutated gene at that locus was responsible for the red (that is, normal) eye color. Responsible? No. Necessary? Yes. The nonmutated gene is necessary but not sufficient for the production of red eye pigment. The *necessity* of the normal gene is proven by the mutant form that leads instead to white eye color. The key species in a natural community is *necessary* for the high species diversity (or, as I suggested for the myxoma virus, its *absence* is necessary). It is futile to attempt to identify the species that is solely *responsible* for the maintenance of observed diversity because no such species exists.

The invasion of Western rangelands by the Klamath weed (Saint John's-wort) and the eventual control of this invasion by the introduction of the leafbeetle, *Chrysolina quadrigemina,* illustrates the difficulty we encounter in attempting to understand community structure. At one time the Klamath weed had overrun and ruined some 2 or 3 million acres or rangeland in California alone; this is a sizable proportion of the total rangeland of that state. In 1946 the leafbeetle was introduced into California in an effort to control this weed. The results were spectacular. The actual advancing front of the beetle population could be seen on the range – grassy areas devoid of weeds behind the narrow advancing front and dense stands of still unattacked weeds in front of it.

The control of the Klamath weed today is virtually complete; its elimination has stopped just short of absolute extermination. The few individual plants remaining are to be found not on rangeland but in shady margins where woodland and rangeland meet. The Klamath weed does not thrive in woodland, but it can survive in the margin where the shade is too deep for the survival of leafbeetle larvae. Without its foodplant, the beetle population is reduced to small numbers of individuals forced to seek out those plants that by chance have come up in sunny areas.

Any naturalist unfamiliar with the history of the Klamath weed over the past half-century or more would draw the following conclusions upon a visit to California rangelands today. (1)   Rangelands are in excellent condition with no obvious threat to their continued and thriving condition. (2)   Scattered here and there in the shady margins of the range there exists a minor weed that plays no role at all in the rangeland community or the range economy. (3)   A relatively rare beetle is seen occasionally on Klamath weeds but the beetle is so rare that it must have a negligible role if any in the ecology of the rangelands. There is no remaining evidence that would tell today's observer that these rare beetles are the key that is necessary for the restriction of the Klamath weed to shady forest edges and that, were this key to be removed, the rangeland of California would once more turn into enormous stands of Klamath weed and be devoid of all valuable forage plants.

The role of keys in the maintenance of diversity in natural communities has been unexplored. The investigation of these factors is to ecology what the investigation of mutant genes was to early geneticists – except that the ecological study must be carried out in the field. The ecologist's understanding of keys and their role in nature makes him nervous about careless, indeed needless, contamination of man's environment. There is reason to believe, for example, that the fourth or fifth levels in food chains – very rare species such as falcons and eagles, which receive only 1/10,000 or 1/100,000 of the energy accumulated in the very first level, are keys to small-mammal variation. Small mammals, as we saw in the case of meadow communities, are keys in turn to plant diversity. In this way, responsibility for much of the natural diversity of man's environment lies with a few high-flying birds of prey, ornithological oddities. And yet, it is precisely in these bird species that the accumulation of DDT is having its worst effects. Falcons destroy their own eggs if the shells are too thin; DDT interferes with the deposition of calcium in egg shells. Throughout the world, populations of falcons and eagles have shown marked declines that parallel the use of DDT in a most striking manner. Consequently, in addition to all the direct effects that DDT may have on insect populations by virtue of the differential susceptibility of the exposed insect species themselves, this insecticide threatens to destroy important keys to the entire diversity of life around us. Tampering with the environment in this manner is an exceedingly dangerous game. It is a fruitless and needless one too. It is done with the avowed purpose of feeding an ever-expanding human population. Until population growth is controlled, makeshift solutions to the ever-increasing demands of human beings will always threaten the very world we would save.

Island Biology
and Public Parks

Oceanic islands intrigue evolutionary biologists. Darwin was tremendously influenced by observations made on the plants and animal life of the Galápagos islands. Each of many forms of life he encountered there reminded him of forms that he had seen earlier on the South American continent several hundred miles away. Nevertheless, the plants and animals that he found living on the various islands in the Galápagos group differed enough from others of the same sort that he could with ease identify the island on which the different specimens had been collected.

The amount of life on remote islands is generally much less than that found on continental land masses. Instead of a wealth of closely related forms living within a short distance of one another, an island may support one or two groups. The dearth of birds, for example, is especially noticeable to a practiced bird watcher. When many bird species live within a small area, a bird's song identifies it as accurately as its color or physical appearance because the different calls serve as recognition signals by which individuals of various species recognize, repel, or attract others of their own kind. On remote islands where the total number of bird species is small, the number of songs is reduced, although to an expert ornithologist the individual variation he can now detect in the calls of one species is surprisingly large.

Within recent years a mathematical theory of island populations has been developed, largely through the efforts of Professor R.H. MacArthur of Princeton and Professor E.O. Wilson of Harvard.* One facet of the theory suggests that the number of species found on an island represents a dynamic equilibrium; although the number may remain relatively constant over extended periods of time, the actual species present on the island may change as some become extinct and others arrive to take their places. The equilibrium number is determined by the relative rates at which colonizers arrive and resident species become extinct. The rate at which colonizers arrive depends, in turn, upon the distance to the nearest large land mass. The extinction rate depends upon the size of the island and its environmental complexity (a feature of an island that is also a function of the island's size).

*R. H. MacArthur and E. O. Wilson, *Island Biogeography* (Princeton, N. J.: Princeton University Press, 1967).

The number of resident species varies with island size in an orderly way. As the size of islands in the South Pacific increase from 1 square mile to 100,000 or more square miles, for example, the number of ant species found on them increases from 5 or less to 100 or more. Similarly, the number of lizard species on West Indian islands increases from 2 or 3 to about 100 as the islands on which they live increase in size from 1 square mile to nearly 50,000 square miles. At a given moment, small islands support very few related species.

Pertinent to the point we wish to make about public parks is the number of species found in circumscribed areas of various sizes that are included within what is a large area. If we use the ant species mentioned above as an example, how many species are found within areas of 1 square mile, 10 square miles, and so forth, within the very largest island that was studied? These numbers are considerably larger than the numbers found on isolated islands of similar areas; 50 ant species are found within a single square mile of the large island, for example, whereas an entire island of 1 square mile would support fewer than 5 species.

The large number of species found within small areas that are parts of larger areas can be ascribed to temporary inhabitants, inhabitants of a sort not found on inaccessible islands. Within large areas there are migrant individuals (wanderers) who are likely to be encountered at any moment at any spot. There are also quasi-permanent residents who live in a given area merely because it is the best spot available to them, not necessarily because it is an excellent (or even an adequate) one. Pairs under such circumstances may be unable to raise young but the parental individuals can be seen and their presence recorded. Each spring for many years I watched a pair of swans build a nest in a marshy inlet on the north shore of Long Island where each year high tides destroyed it before the eggs had hatched. Other, better areas were occupied by other pairs of swans that were uniformly successful in raising cygnets (most of which were later victims of snapping turtles).

Island biogeography is particularly relevant to the encroachment of man on his natural environment, to the spread of suburbs and housing developments into naturally wooded or rural areas. The effect of such encroachment has been described by Chekhov speaking through Astrov, the country doctor in the play "Uncle Vania":*

ASTROV: I don't suppose this will interest you.
YELIENA: Why not? It's true I don't know the country, but I've read a great deal.
ASTROV: I have a table of my own here . . . in Ivan Petrovich's room. When I'm completely tired out − to the point of utter stupefaction − I leave everything and escape here to amuse myself for an hour or two with this thing . . . Ivan Petrovich and Sophia Alexandrov click away at their counting

*Reprinted from *Uncle Vania* by Anton Chekhov, translated by Elisaveta Fen, by permission of the publisher, Penguin Books Ltd.

beads, and I sit beside them at my table and mess about with my paints – and I'm warm and quiet, and the cricket chirps. But I don't allow myself this pleasure very often, just once a month . . . (*Pointing to the chart.*) Now look at this. It's a map of our district as it was fifty years ago. The dark and light green stand for forest; half of the whole area was covered with forest. Where there's a sort of network of red over the green, elks and wild goats were common . . . I show both the flora and the fauna here. This lake was the home of swans, geese, and ducks. There was 'a power' of birds, as the old people say, of all kinds, no end of them; they used to fly about in clouds. Besides the villages and hamlets, you can see all sorts of small settlements scattered here and there – little farms, monasteries, water mills . . . Cattle and horses were numerous – that's shown by the blue colour. For instance, there's a lot of blue in this part of the region. There was any number of horses here, and every homestead had three on the average. (*A pause.*) Now let us look lower down. This is how it was twenty-five years ago. Already only a third of the area is under forest. The goats have disappeared, but there are still some elks. The green and the blue colours are paler, and so on. Now look at the third section, the map showing the district as it is now. There's still some green here and there, but it's not continuous – it's in patches. The elks, swans, and woodgrouse have all disappeared. There's not a trace of the small farms and monasteries and mills that were there before. . . . In general, it's an unmistakable picture of gradual decay which will obviously be completed in another ten or fifteen years. You may say that it is the influence of civilization – that the old way of life naturally had to give place to the new. Yes, I would agree – if on the site of these ruined forests there were now roads and railways, if there were workshops, and factories, and schools. Then the people would be healthier, better off, and better educated – but there's nothing of the sort here! There are still the same swamps and mosquitoes, the same absence of roads, and the dire poverty, and typhus, and diptheria, and fires. Here we have a picture of decay due to an insupportable struggle for existence, it is decay caused by inertia, by ignorance, by utter irresponsibility – as when a sick, hungry, shivering man, simply to save what is left of his life and to protect his children instinctively, unconsciously clutches at anything which will satisfy his hunger and keep him warm, and in doing so destroys everything, without a thought for tomorrow. . . . Already practically everything has been destroyed, but nothing has been created to take its place. (*Coldly.*) I see from your expression that it doesn't interest you.

YELIENA: But I understand so little about all this. . . .

ASTROV: There's nothing to understand: your're just not interested.

YELIENA: To be quite frank, my mind was on something else.

One point brought out explicitly by Astrov's monologue is that small isolated patches of woodland do not support the variety of wildlife that the larger more continuous woodlands once did. In part, this fact merely reflects the need of large animals to have a suitable range from which to obtain their food. In the case of grazing animals, a suitable range can be understood in terms of acres per head. A pet horse cannot be supported by the grass growing on the lawn of a city lot, nor can a herd of wild goats maintain itself in a small patch of forest and meadow. Correspondingly, a carnivore needs ample space within which its herbivore prey might make a living; again it turns out that only one carnivore can be supported within a certain area. There is, in short, a maximum population density for each type of animal species.

In addition to the sheer inability of small isolated woodlands to support certain forms of life, the number of species is further reduced because the accidental loss of a species from one woodland cannot be restored easily by migration from another. An empty but otherwise suitable "home" is rather quickly found and occupied in a large continuous area where wanderers are constantly moving about in active search of just such places. An equally desirable "home" may go empty for long periods in a small patch of woodland because the species does not exist there and wanderers cannot find their way to that isolated patch. Small patches of woodland scattered here and there over a wide area are comparable to isolated oceanic islands; the flora and fauna they support are grossly impoverished.

The lesson that island ecology teaches us is a simple one to understand. Natural habitats that are set aside as public parks and wildlife refuges will retain and support the greatest variety of wildlife if they are large continous areas rather than isolated, neighborhood parks. Mini-parks have a place, a wonderful place, within the city; there the weary pedestrian seeks only a cool green spot within which to rest. In the country, where public parks are expected to preserve many types of plant and animal life, their individual sizes must be appropriate for the job they are expected to do. If small neighborhood parks are desirable, they must be located in a pattern such that interconnecting greenbelts can be built to link them into what are effectively much larger units. It is through these belts that the migrants will come, migrants from one park that will replace the accidental losses of another. And where can the land be found from which greenbelts can be made? A great many persons have suggested that abandoned and neglected railroad rights of way would serve admirably for this purpose.

The Estuary:
One of Nature's Keystones

Venice sits on the common estuary of three rivers: the Piave, the Brenta, and the Sile. It is not a normal estuary, however, because many centuries ago the Venetians diverted the main river channels and joined the sandbars (*lidi*) in order to make a large, permanent, enclosed lagoon. Had they not done so, the silt brought down by the rivers would have accumulated behind the outer sandbars; left to itself, the estuary, in time, would have become a coastal salt marsh, an area of mud and sand flats interspersed between meandering streams filled with brackish (half-fresh, half-salt) water.

The coastal salt marsh that would have formed on the site where Venice now stands was prevented from doing so. The prevention has not been complete, for the lagoon is shallow and at low tide islands appear that are submerged again when the tide is high. On the floor of the lagoon stream channels meander; these can be identified on hydrographic charts. The salt marsh has always been a favorite target for land fills. Ravenna, for example, occupies its estuary not as a city of canals as Venice does but as a city built on fill. The Netherlands is an entire country largely created from the shallow seas of the coastal area; in this case, the water has been excluded systematically by man-made dikes rather than replaced by land fill. Nearly all large coastal cities of the United States have their nearby marshland to which rubbish is trucked to be burned and to accumulate; eventually these are the sites of new housing developments that are needed to absorb the expanding population.

The topic of this essay is the ecological and economical impact wrought by the destruction of salt marshes. The title refers to the estuary as a keystone. It is just that. It supports a great deal more than the dozen or two shabby men who can be seen at low tide digging for clams and who appear to be the only persons saddened when the zoning board chooses the salt marsh as the site for the new city dump. An estuary with the properties we recognize as estuarine is built where a river enters the sea. The water of the river carries soil and various nutrients that it has collected from distant inland points. Where it enters the sea, conflicting currents, eddies and other turbulences, and the varied sizes and weights of suspended particles cause the estuary to grow as a series of marginal mud banks and sandbars. Algae and other plants that can tolerate salt water take root on these banks. More sediment is trapped, the banks grow higher, other plants move in, and soon a series of islands form – some permanent, some intertidal. Throughout the whole marshy area run meandering streams that fuse into channels and at last break through to the sea. The Lido of Venice is one of the early sandbars. Originally there were four *lidi* separated by five channels; the early Venetians made these four *lidi* into two (of which one is the Lido of tourist fame) after the main river channels had been diverted.

The amount of nutrient material in an estuary and the amount of life it supports is fantastic. The slow currents, the shallow depth of the water, the ease of aeration, and the concealment offered by a profusion of aquatic plants make the estuary (now a salt marsh) a haven for a host of small marine organisms. It is no accident that mussels, clams, oysters, prawn, and other sessile and near-sessile animals thrive in these marshy flats, for the water they imbibe in order to feed (an oyster draws in 20 quarts of seawater each hour) is a thin soup of small swimming organisms.

The estuary is a keystone for marine life because of its position in the food web of the sea. Of the total harvest of marine fish (weakfish, bluefish, mullet, menhadens and others), 90 per cent is harvested from the shallow waters over the continental shelf, the submarine plain that extends 200 miles or more from the coastline of every continent. Of the fish taken from the region of the continental shelf, two-thirds spend their juvenile stages in estuaries; 60 per cent of all fish harvested from the sea start life as small fish darting in and out among the plants of estuaries and tidal marshes. Filling these marshes with rubbish and dirt in preparation for the apartment houses of 10 years hence or in the construction of the new airport that promises to bring so many benefits (such as increasing the demand for new apartment houses) does more than displace a few eccentric clam diggers; it displaces an inordinately large segment of our annual food production. Tidal areas are areas that deserve to be protected vigorously. The fishing and canning industries should look askance at any attempt to destroy what are in all truth the seedbeds or germination flats of their livelihood. Those responsible for the health of the nation should guard more zealously the ultimate sources of our food — fertile farmland and tidal salt marshes.

The estuary is a keystone, too, in the maintenance of water fowl. The overwintering feeding grounds of geese, ducks, swans, rails, and other birds have traditionally been the tidal estuaries, the wetlands, of the Atlantic and Gulf coasts. Destroy winter feeding grounds, of course, and the tremendous annual migrations of these birds to the North will cease — as will the whole superstructure of hunting and camping as an outdoor sport for millions.

Because the estuary lies at the mouth of a river whose drainage basin may include thousands of square miles, it is at the mercy of a great many persons who live far from it and a great many activities that take place hundreds of miles away. Because 200 pounds of an insecticide were spilled accidentally into the Rhine River in West Germany, tons of fish were killed and for nearly a week Rotterdam lived under emergency water control measures. Rotterdam, you see, occupies the estuary of the Rhine. Either accidents or indifference can lead to the biological demise of any salt marsh. A paper mill hundreds of miles upstream can smother the floor of a marsh with chemicals and pulp fibers; once the bottom dwellers are dead, the shellfish and juvenile stages of marine fish starve too. Runoff from miles of agricultural land can carry enough insecticide into a river to destroy the estuary that lies at its mouth. Waste chemicals from refineries and other chemical processing plants can do the same. The famous oyster beds on the south shore of Long Island were all but destroyed by the too-rich drainage from the equally famous Long Island duck farms.

In this topsy-turvy house of cards that man has undertaken to build and inhabit, perhaps few pieces are superficially as dull or less attractive than estuarine marshes — flat, insect-infested, and drab. To those who know them well, however, these represent key pieces holding steady many seemingly more important and more impressive ones. These marshes are the seedbeds of the sea; they also serve as the feed trough for migratory water fowl. Their wholesale destruction promises to initiate a series of tremendous and unpleasant ramifications with man, as usual, the eventual loser.

On the Stability
of Biological Systems

Biological systems as a rule are stable systems. The metabolic activities taking place within cells are controlled at a multitude of crucial points by feedback mechanisms that turn off unnecessary reactions and turn on necessary ones. The terminal chemical compound produced by a sequence of enzymatic reactions is often the substance that is capable of inactivating the first enzyme of the sequence. Thus, once the sequence has operated so that the end product exists, the end product itself turns off the entire sequence. By analogy, once a furnace has run for a short time, its end product − heat − activates the thermostat that turns off the furnace.

The body functions similarly, through the interplay of a series of regulatory, or *homeostatic*, devices. A substantial loss of blood causes a drop in blood pressure; this in turn stimulates the release of adrenalin, which promotes rapid clotting of blood. The ease with which a sample of blood will clot increases with the amount of blood lost before the sample was taken. Bleeding, if not too severe, is a self-correcting process.

Within the world at large, natural communities tend to be relatively stable in their composition. An increase in the numbers of one small mammal relative to others of comparable size and similar habitat is halted and reversed by the growing ease with which disease spreads from individual to individual, or by the greater susceptibility of the excess individuals to capture by local predators.

The self-correcting abilities of living systems are familiar ones, and so many people believe that it is nearly impossible to do serious or permanent harm to our environment or to the ecology of a region. In this essay I discuss the meaning of stability as it relates to biological systems. My theme is that not all stable systems are equally resistant to disturbance and that the stability of some systems is actually quite fragile. Thus, what we do *not* know about the stability of our environment may at times hurt us severely.

The stability of a dynamic system lies not in its refusal to budge when disturbed (as the giant pyramids of Egypt might do) but in its tendency to return to an initial state after being displaced (as a pendulum might). This is the essence of the stability of dynamic systems. The displacement, however, must be specified as "slight." A brick has three stable positions: on its face, side, or end. A slight displacement is reflected in a slight tilt of the brick but, if the displacement is truly slight, the brick will return to its original stable position after the displacing force is withdrawn. A sizable force, on the contrary, will topple a brick from one position to another – from balancing on end, for example, to lying on its face.

Cyclic phenomena have their stabilities too; the boom-and-bust-cycles of population growth in the case of many animals are most often stable cycles. Change per se is not a sign of instability. Stability in the case of cyclic changes requires that the cycle run its course and return to its initial starting point; stability in the face of small displacements means that at any point in the boom-and-bust cycle the population is permitted to deviate somewhat from the "normal" course of events without leaving the cyclic pattern. Raceway drivers who leave the track momentarily and run onto the grass next to the racetrack only to return and continue on their way illustrate the stability of cyclic phenomena.

Raceway drivers who leave the track and crash into the retaining wall do not return to the race; they represent otherwise stable, cyclic systems that have been displaced too far. Laboratory populations of *Drosophila* flies oscillate somewhat in size; the chance extinction of such populations very often follows an exceptionally large burst in population numbers. These populations are lost because of wholesale starvation among the tremendous numbers of larvae of the post-boom generation. The few adult flies of this starved generation that do survive are invariably small and sterile.

Until recently man has known remarkably little about the dynamics of equilibrium conditions within cells, within whole organisms, or within the world at large. Isolated facts he has known; the interrelations of facts have very often been obscure. Certain biochemical compounds and the pathways by which they are synthesized have been known for a century or more. Feedback inhibition of synthetic pathways has been known for one or two decades. The changes in molecular structure that accompany the change from an active to an inactive form of an enzyme have been understood even in part only very recently.

Poorly understood physiological or neurological equilibria will reappear as a rather serious matter when we consider birth control practices and when we discuss drugs and drug addiction. Ignorant as we may be about certain aspects of our internal physiology, the depth of our ignorance in such matters is far less than that which we – scientists and nonscientists alike – exhibit in respect to our external environment. Perhaps the examples that follow will instill in us a sense of humility in matters that concern indiscriminate, often needless, tampering with the environment.

The 16-armed, spiny sea star (the "crown of thorns" starfish) has been indicted for the destruction of coral in the Red Sea (reported in 1963), along the Great Barrier Reef in Australia (reported in 1968), and near the island of Guam (reported in 1969). Now, the ordinary, beach-variety starfish is an extremely efficient clam and oyster eater; the single muscle that holds shut the two half-shells of a clam is no match for the many-armed, many-muscled predator that exerts a continuous pulling action on the shell by means of its suction-cupped "feet." When the clam reluctantly opens its shell, the starfish floods its victim with digestive juices, everts its stomach, and eventually absorbs the already digested clam. The spiny sea star is a large starfish; it everts its stomach and digests the small polyps that are responsible for the formation and continued maintenance of coral reefs. Once the polyps have been killed and eaten, the reef becomes merely a chalk-like edifice that is subject to rapid erosion by water currents and wave action.

Ordinarily, the spiny sea star is a rare animal; on infested coral reefs, however, an alert diver can spot as many as 100 in 10 minutes. Foraging "herds" occur near Guam in bands 10 yards wide and a mile or two long within which there may be an animal in every square yard; 50,000 or more adult individuals may be found in a small region of the island's coastline.

The damage these starfish are capable of doing is tremendous. Many — perhaps most — of the small islands of the South Pacific exist only because of encircling coral reefs that protect them from the waves. Destroy the reefs and the islands will disappear. When we talk of islands disappearing, we are talking of vast regions of the South Pacific where millions of persons live; for these persons the destruction of coral reefs threatens not only their food supply but also the very sand under their feet.

Great as the damage caused by this infestation of starfish could prove to be, no one is sure why there has been an outbreak. Are skin divers selectively collecting one or more of the starfish's natural predators? Perhaps. Some of these predators have shells that are prized by private collectors, and skin divers have been collecting and selling them in large numbers. Blasting? Dredging? Testing nuclear devices? Perhaps. On the other hand, the young of the spiny starfish are an easy prey to many forms of marine life; among other predators are the polyps that construct coral reefs. The burst of adults now seen may reflect some sort of hidden harm done to organisms that might otherwise devour the young at an early, helpless stage. Is today's starfish population explosion the boom portion of a boom-and-bust cycle that repeats itself at rare intervals? Perhaps. Truthfully, no one knows why the infestation has occurred, and no one knows precisely how to bring it under control.

Forest fires, as all of Smokey the Bear's many friends know, destroy millions of acres of valuable timberland each year. The dry forests of the Far West represent an especially vulnerable area. As one raised in the East, I was amazed when I first dug into the needles on the floor of a ponderosa pine forest in Southern California (a forest now largely destroyed by automobile exhaust from Los Angeles); pine needles had accumulated to a depth of a foot or more but these were as dry, perhaps drier, than the hay in the loft of a barn in New York or Pennsylvania. Small wonder that a cigarette tossed carelessly from a car window can start a fire that will burn out of control through miles of Southern California forests.

After a generation in which the absolute elimination of forest fires has been one of its goals, the Bureau of Land Management is having second thoughts. Not all fires are manmade; many are started by lightning. Fires have been recurrent events in forests for thousands of year; they occurred even before man's arrival. Analyses of the growth rings of mature ponderosa pine trees have shown that these trees in past ages have lived through frequent fires that burned only the bases of the tree trunks. Slowly a coherent picture of forest ecology is emerging. Ponderosa pines are resistant to periodic brush and litter fires that burn the trash and young seedlings on the forest floor; they are destroyed by fires that occur so infrequently that many medium-sized trees have time to grow up during the fire-free interval. The medium-sized trees serve as a springboard by means of which a harmless fire in the low underbrush becomes a devastating "crown" fire that rages through the tops of mature trees. With the new appreciation of the role of frequent small fires in forest ecology comes the further appreciation of the complex relationships between the inhabitants of forested regions and between the inhabitants and the forest that is their home.

Whether they are made thoughtlessly or with seemingly careful logic, man's decisions often create disturbances that, like ripples on the surface of a pond, spread throughout the world in which he lives. The wholesale elimination of eagles, falcons, and condors may represent the loss of just a few birds to many people, but to others they represent the loss of keystones that lend stability to the natural environment — stability to the relative proportions of small mammals and, through these mammals, to the number and distribution of plant species in fields and meadows. The inhabitants of coral reefs may have appeared to be fair game to scuba divers who for so long peered, photographed, and speared them virtually at will. The destruction of coral reefs (and of the sandy atolls they protect) by the spiny starfish remind us, however, that marine communities also have their delicately balanced equilibria. Disharmonies that accompany the near elimination of forest fires remind us that organisms and natural communities through millenia of natural selection have adapted to existing conditions; these adaptations need not be wholly functional in man's "improved" version of environmental ecology.

Where does all this leave us? For the most part, it leaves us with a sense of wonder at the depths of our ignorance in respect to the interactions of complex systems. In addition, it leaves us with a sense of distrust (and distaste) for those who so glibly promote their own interests while pooh-poohing the possibility that serious, unwanted side effects may be in the offing. It leaves us, I believe, with an appreciation for the place of healthy skepticism in our society and for the important role played by the intelligent and articulate skeptic — the gadfly of those truly dangerous individuals who possess power but lack both foresight and wisdom.

Reaching a Simple Decision

To take or not to take an umbrella when leaving the house is a daily problem for commuters, shoppers, students, and other persons. In most instances it is a problem solved by intuition; nevertheless, it can be analyzed in a rational manner.

The problem consists of two components: whether or not I take my umbrella and whether or not it will rain. The four possible combinations of these alternatives can be represented by a small, four-celled checkerboard.

		Take umbrella?	
		Yes	No
Rain?	Yes		
	No		

To reach a decision, I assign a value to each of the four combinations. If it does not rain and I am not carrying my umbrella, I am happy; I assign this combination the value 100. If it rains and I have my umbrella, I am as happy as I can be under rainy circumstances and so this box is also assigned the value 100. The least happy situation is being caught in the rain with no umbrella; this combination is assigned the value 0. Finally, the inconvenience of carrying an umbrella unnecessarily, while not as bad as getting soaked, calls for an intermediate value; I assign it the value 20. The checkerboard now looks like this:

		Take umbrella?	
		Yes	No
Rain?	Yes	100	0
	No	20	100

The final step in the decision-making procedure is the evaluation of the

250

likelihood of rain. This evaluation can be made by studying the sky intently with one hand outstretched, palm up, as if in supplication or, in certain localities, by reading the daily paper. The weather report, for example, may say that the probability of precipitation is 30 per cent; in that case, 0.70 and 0.30 can be used as weighting factors in arriving at a decision:

| | | Weighting factor | Take umbrella? | |
			Yes	No
Rain?	Yes	0.30	30	0
	No	0.70	14	70
	Sum		44	70

Satisfied that the values assigned to the four combinations and the weighting factors are the best possible ones, I decide not to take my umbrella. Others, less sensitive to the inconvenience of carrying an umbrella on a clear day, might use other values and reach the opposite decision. Similarly, my own decision is dependent upon the weatherman's forecast. My decision to leave the umbrella at home, given the described circumstances, is a *proper* decision. Later in the day, however, I might find myself caught in a downpour with no protection and, thus, realize that my decision was a *bad* or *erroneous* one even though properly arrived at.

Because decisions, even when properly made, can err, the routine decision-making process outlined here is most useful in arriving at trivial decisions, which if wrong lead to remediable, not irreparable, consequences. Decisions that may lead to irreparable harm, such as a decision to play Russian roulette, as the following essay emphasizes, must be made by extraordinarily thoughtful, *not* routine, procedures.

Brinksmanship:
The Limits of Game Theory

Games and decisions fall within the realm of formal mathematical analysis. Certain information is fed into standard equations, some calculations are performed, and the most profitable course of action is identified. In effect, a decision is made. At times the decision involves what is least obnoxious rather than most profitable; the need to dump into the sea 12,000 rockets armed with deadly nerve gas arose in the summer of 1970 following a series of earlier blunders whose consequences could not be undone. Whatever the term, however, the mathematical analysis leads to a decision; it governs a course of action.

The success of decision theory has blinded many persons to its limitations; mathematical analyses have a way of lulling the unwary. Assumptions are made for the sake of simplifying calculations; at the time they are made, these assumptions are justified as *reasonable*. Later, should the results of a given analysis be challenged, the same assumptions are paraded as actualities. I intend to question the use of routine decision theory in matters where damage once done is enormous and irreparable. *Brinksmanship*, in this essay, is the practice of using game theory under boundary conditions where its basic assumptions are no longer applicable; it is the art of rationalizing Russian roulette by logic designed for pocket billiards.

The approximations and simplifications that are inherent in any applied mathematics limit the extent to which subsequent calculations can be used in understanding the real world. In discussing gene frequency changes under the influence of natural selection, for example, I must be very careful when talking with a plant or animal breeder. In my work I deal with genes whose frequencies are very low and, consequently, I employ one set of algebraic equations. Plant and animal breeders, on the other hand, are interested in genes whose frequencies are substantial and, for that reason, they use a different set. The equations that I use are derived from theirs by omitting terms that I consider negligible but which they cannot afford to discard. The equations that we both use and understand, however, are based on assumptions that make little sense to a biochemical geneticist or a molecular biologist.

Decision-making encounters similar restrictions. The sad fact is that formal decision-making is useful only in making decisions of a trivial nature — those whose consequences are reversible or, if they happen to be irreversible, where subsequent action can nevertheless rectify an earlier harm. This claim applies even to such matters as Japan's decision to bomb Pearl Harbor and thereby enter World War II and various decisions made by U.S. naval commanders during the Battle of Coral Sea; the consequences of erroneous decisions in each instance were not irreversible. The formalized aspects of decision-making falter when an act capable of inflicting enormous *(infinite,* in lay terms) harm has an exceedingly small probability of occurring ("not one chance in a million" or "not one chance in a billion"; although these expressions are used interchangeably, if the first means annually, then the second means twice since the birth of Christ). In such cases, routine calculations based on assumptions, approximations, and conveniences that have proven useful in other, minor affairs are not to be trusted.

An excellent illustration of an undertaking of the sort where routine decision-making techniques are likely to lead to gross errors — to irreparable damage — is the construction of multimegakilowatt nuclear generators. This is not to say that such generators should never be built; I am simply pointing out that their construction is part of a game that is played *once.* My claim is that routine decision-making techniques are inadequate because, in the case of an accident, each power plant of this sort could make a large segment of the nation uninhabitable for decades; the chance that such an accident will occur, however, is very small. These are what I have called boundary conditions; consequently, techniques and attitudes suitable for trivial decisions in which both the values and the probabilities assigned to various possible outcomes are roughly equal (or at least not more different, for example, than 100 to 1) are inadequate for coping with this more serious type of problem.

Should the chain reaction of a nuclear power plant run rampant by accident, radioactive debris could be scattered over considerable distances. Farmland would be rendered useless and uninhabitable. Nearby and downwind cities would require quasi-permanent evacuation. The economic consequences of such disasters have been estimated.* In the case of an accident a 1-million kilowatt reactor situated in the Midwest could cause damages estimated at nearly 1 billion dollars at preinflationary prices. About one-half of this amount would be needed to purchase or lease the land and buildings that would be unfit for use for two decades or more. Under present federal legislation the total guaranteed liability in case of such an accident is limited to just this lower amount: $500,000,000.

The rationale for continuing with the construction of such potentially destructive power plants rests, of course, on several factors: (1) their need in meeting growing energy demands, (2) the profit that each will make for its owners — a profit that is reasonably large in comparison with the maximum legal liability, and (3) the extremely small probability that a devastating accident will

*C. Rogers McCullough, ed., *Safety Aspects of Nuclear Reactors* (New York: Van Nostrand Reinhold Company, 1957).

occur. Under less cosmic circumstances, a rationale based so soundly would be unassailable. On the scale of megakilowatt power plants, the rationale is vulnerable from all three aspects.

The nation's energy demands have doubled every 9 or 10 years for the past half-century. Is it not true, therefore, that we must be prepared for even greater demands? Not necessarily! At some time a thorough inventory of the nation's use of energy and the reasons behind each additional request must be made because a perpetual doubling cannot continue. Now seems to be a reasonable moment to start. How much energy is used to maintain the one-way flow of materials in our lives: disposable bottles, phosphates in detergents, phosphorus and nitrogen in fertilizer, "disposable" automobiles? What about planned obsolescence generally? In a crunch, energy requirements can be met by decreasing demands as well as by increasing energy production. For a number of reasons it is obvious that the crunch is coming; the present transition stage between conventional and nuclear power sources offers an opportunity for taking stock.

The profits of a nuclear power plant promise to exceed its legal liability. Is that not reason enough to proceed with its construction? Not as I see it. Legal restrictions on liability do not concern me; I prefer to discuss *real* liability. Two decades ago, engineers calculated that in case of an accident one-half billion dollars would be needed to purchase the contaminated land surrounding a Midwestern nuclear power station. Inflation has made this estimate obsolete; nevertheless, the legal limit has been set by law at this inadequate level. What does it mean, however, to buy a half-billion dollars worth of property at a random moment in time – a time literally set by accident? Ordinarily a person sells his home or business because he is moving to a new location or, if he is remaining in the same neighborhood, because he has had an opportunity to make a new and better purchase nearby. A nuclear accident removes these logical aspects of buying and selling property. Thousands of persons would be told outright that they must leave the disaster area; there would be no provision for resettlement, for creating new employment, or for patching torn lives. There would be no liability payment above that needed for the condemnation of property; no funds would remain to help defray medical expenses, moving bills, or living costs during the period of resettlement. Once it appears that these human aspects of the possible tragedy have been totally neglected, one also suspects that the cost of land – because of the large quantities involved – has been computed at wholesale prices.

At no time has the removal of farmland from useful agricultural production been included in the estimation of damages from a nuclear accident. As the population grows and farmland becomes ever more precious, the nuclear stations that are proposed can effectively destroy tens of thousands of acres instantaneously. Furthermore, to my knowledge, no consideration has been given to the fate of the radioactive contaminants in the soil. Do they run off into nearby streams, lakes, and rivers – thus joining DDT, dieldrin, and mercury as major pollutants? Do the radioactive elements start an inexorable journey through the web of life by moving from soil to plants to herbivores to carnivores and man — increasing in concentration with each step? I, for one, am not at all convinced that the liability for nuclear disasters has been evaluated correctly.

What does liability matter if the risk of an accident is so extremely low? Nuclear reactors are built so that they cannot fail. These comments represent the contributions of a safety engineer to the decision-making process. Can such arguments be trusted? Again, I have reservations. How much of this expressed confidence represents blind faith in the widely touted American know-how? How much of it is based on the traditional assumption that should a mistake occur it can be corrected on the next try? "Back to the old drawing board" does not apply to nuclear reactors. I would suggest that these reactors have not been remarkably fault-free; Sheldon Novick has documented the accidents that have plagued these devices from the outset.*

Enough has gone astray in modern technology to suggest that techno-logical planners are not exceptionally gifted in either matters of foresight or the anticipation of novel or unexpected events. The continual beeping of one of our early space satellites long after its assigned task was over showed that no one had anticipated turning it off. The televised actions of the astronauts during and after the first moon voyage when strict sterile techniques were supposedly in force in order to test for life on the moon and to prevent the possible contamination of the earth by extraterrestrial organisms were appalling to anyone trained in microbiology: Moon dust that was to be tested for extra-terrestrial life was obtained from an area thoroughly marred by the astronauts' own footprints; upon entering the isolation chamber after their return, at least one astronaut eased himself into the rather small doorway by grasping the outside of the chamber. The list of blunders committed by the Army in its development and subsequent disposal of an arsenal of nerve-gas weapons is enough to make any but the most avid technologist lose faith.** At Lausanne, Switzerland, a nuclear reactor had to be destroyed and the Alpine cave in which it was located sealed up — thus showing that the chance of a nuclear accident is not zero. The Swiss reactor, incidentally, was originally intended for construction in the city of Lausanne itself, but the local inhabitants, with justifiable concern, insisted on the more remote location where the accident occurred.

The effectiveness of routine decision-making procedures breaks down under boundary conditions, that is, when dangers become incalculably high and risks inestimably low. This is the realm within which decision making becomes brinksmanship. Decisions, even under these conditions, must be made, never-

*Sheldon Novick, *The Careless Atom* (Boston: Houghton Mifflin Company, 1969).
**See Tom Wicker, "The Nerve Gas Affair," *The New York Times* (Aug. 12, 1970). There are also many other news releases available of that time.

theless, and not all can be biased toward extreme caution. It would be fatal, however, if public apathy were to lull decision makers into forgetting the special and hazardous nature of brinksmanship. Decision makers must not be permitted to justify the construction of one nuclear reactor merely by an earlier one's existence nor should they be encouraged to keep pace with growing energy demands rather than question the demands themselves. Public skepticism should confront plans for the construction of *each* nuclear power plant; furthermore, authorized construction of *each* one should proceed only after an extraordinary (*not* routine) examination of all pertinent facts. Long-lived radioactive material permits us to play the nuclear-power game but once; there is no "next time."

Addendum

The demand for electrical energy in the United States has doubled every nine or ten years for nearly a half century. Much of the nation's thermal pollution, much of its air pollution, and the threat of pollution by radioactive wastes stem from this enormous demand for power. In the accompanying essay I have suggested that an inventory of energy needs should be taken. The following statements outline the energy demands posed by one small facet of an activity that many Americans seem to regard as "normal": waging war in Indochina.

Item: *New York Times*; Sunday, March 14, 1971 (page 1, Section IV) Total tonnage of bombs dropped in Indochina war: 5,700,000
Item: *New York Times*; Saturday, March 20, 1971 (page 39) (Headline) 2 GENERATING UNITS PLANNED

1. The manufacture of 233 pounds of TNT requires 189 pounds of nitric acid and 98 pounds of benzene. Therefore, 6 million tons of TNT require 4.8 million tons of nitric acid (HNO_3).
2. The manufacture of 504 pounds of HNO_3 requires 204 pounds of ammonia. Therefore, 5 million tons of HNO_3 requires 2 million tons of ammonia (NH_3).
3. The manufacture of one ton of NH_3 by nitrogen fixation requires 65,000 kilowatthours of electricity. Therefore, 2 million tons of NH_3 require 1.3×10^{11} kilowatthours.
4. The two plants mentioned in the New York Times have a combined capacity of 800,000 kilowatts.
5. These plants would require (1.3×10^{11} kilowatthours) ÷ (8×10^5 kilowatts) or 160,000 hours to produce 2 million tons of ammonia. 160,000 hours equals approximately 18 years.
6. Consequently, to produce ammonia (and just the ammonia) at the rate needed to maintain the production of TNT for the Indochina war, at least six of these newsworthy generators would be required.

Appendix A

The following essays in volumes II and III of *Essays in Social Biology* discuss matters related to the sections of this volume:

People and Their Needs
>The arithmetic of epidemics   III

The Environment
>Radiation in research and industry   II
>On the glib assurance   III

Ecology
>On the evolution of insular forms   II
>Meeting conflicting demands   II
>The flow of information   III

Appendix B

Table of Contents: Volume II

SECTION ONE: GENETICS

Introduction

GROWING UP IN THE PHAGE GROUP, *James D. Watson*

THE DOUBLE HELIX, *James D. Watson*

DNA: Blueprint for Life: An Essay in Three Parts
1. *Self-replication as the basis for life*
2. *The control of physiological processes*
3. *Heredity and individual development*

The Central Dogma: The Preservation of Simplicity

Genetic Engineering: The Promise Of Things to Come

The Impact of Genetic Disorders on Society

Genetics under Stalin: The Suppression of a Science

SECTION TWO: EVOLUTION

Introduction

THE VOYAGE OF THE BEAGLE, *Charles Darwin*

On the Evolution of Insular Forms

Natural Selection: Directional and Stabilizing

Meeting Conflicting Demands

Instant Speciation

On Modern Genetics and Evolution

The Evolution of Ideas and Beliefs

Technology and the Future of Man

SECTION THREE: RACE

Introduction

REBELLION IN THE BACKLANDS, *Euclides da Cunha*

On the Biology of Race: An Essay in Three Parts
   1. Genetic variation within local populations
   2. Genetic variation between local populations
   3. Races and racial differences

The Role of Labels in Describing and Understanding Evolving Systems

A Potpourri of Racial Nonsense

Race and IQ: A Critique of Existing Data

SECTION FOUR: RADIATION BIOLOGY

Introduction

HIROSHIMA, *John Hersey*

ATOMS AND RADIATIONS, *Bruce Wallace and Th. Dobzhansky*

Radiation in Research and Industry

The Immediate Effects of Radiation Exposure

Radiation and the Unborn Child

Radiation and Cancer

Radiation and Future Generations

Table of Contents: Volume III

SECTION ONE: DISEASE

Introduction

A JOURNAL OF THE PLAGUE YEAR, *Daniel Defoe*

FROM THE FARM OF BITTERNESS, *James A. Michener*

MY LIFE AS A YOUNG GIRL, *Helena Morley*

THE DEATH OF IVAN ILYCH, *Leo Tolstoy*

A NATURAL HISTORY OF THE DEAD, *Ernest Hemingway*

The Arithmetic of Epidemics

Epidemics and Epidemic Diseases

The Biology of some Diagnostic Symptoms

On Natural Selection

On the Coadaptation of Hosts and Parasites

Some Letters Home

Causes of Mortality in Man: Then and Now

The Biology of Cancer

Death as a Biological Phenomenon

On the Dignity of Dying

Suicide and Man's Right to Die

SECTION TWO: SEX

Introduction

THEME IN SHADOW AND GOLD, *Carl Sandburg*

On the Spread of Venereal Disease

On the Role of Sex in Nature

Cousin Marriages: An Essay in Three Parts
   1. Mutant genes in populations
   2. Children of cousin marriages
   3. Genetic counseling

The Biology of Birth Control: An Essay in Three Parts
   1. Hormones: Internal messengers
   2. The female reproductive cycle
   3. Artificial birth control

SECTION THREE: COMMUNICATION

Introduction

FROM THE INLAND SEA, *James A. Michener*

THE EMPTY CANVAS, *Alberto Moravia*

THE WEB OF FRIENDSHIP, *William H. Whyte, Jr.*

Why Communicate?

Systems for Communicating

The Accuracy of Messages that are Sent and Received: Two Case Studies

On the Natural History of Student Unrest

On the Role of Rituals in Communication

On the Glib Assurance

On the Dullness of Scientific Writing

The Flow of Information

SECTION FOUR: BEHAVIOR

Introduction

OF THE INCONSISTENCY OF OUR ACTIONS, *Montaigne*

ON BEING FINITE, *J.B.S. Haldane*

JOBS, *Paul Goodman*

ADOLESCENT DRUG DEPENDENCE: AN EXPRESSION OF AN URGE, *Roy Goulding*

The Nervous System and its Organization: An Essay in Five Parts
1. *The gross structure of the brain*
2. *The circuitry of the central nervous system*
3. *Memory: The storage and recall of information*
4. *The primitive brain*
5. *The individual nerve cell: To fire or not to fire*

On the Inheritance of Behavior

On Kidding Ourselves

On Escaping Reality

On Obedience and Conformity

Epilogue

Index

An alphabetical listing (based on **KEY WORDS** of their titles) of the essays of this Volume.